Solutions Manual for

Geometry: A High School Course
by S. Lang and G. Murrow

Philip Carlson

Solutions Manual for

Geometry:
A High School Course
by S. Lang and G. Murrow

With 100 Figures

Springer-Verlag
New York Berlin Heidelberg London Paris
Tokyo Hong Kong Barcelona Budapest

Philip Carlson
General College
University of Minnesota
Minneapolis, MN 55455-0434
USA

Mathematics Subject Classification (1991): 02/04, 02/05

Library of Congress Cataloging-in-Publication Data
Carlson, Philip.
    Solutions manual for *geometry: a high school course* by S. Lang
and G. Murrow/Philip Carlson.
       p. cm.
    Includes bibliographical references.
    ISBN 0-387-94181-9
    1. Geometry, Plane—Problems, exercises, etc.   I. Title.
QA459.C24   1994
516.2—dc20   93-38093

Printed on acid-free paper.

Production managed by Francine McNeill; manufacturing supervised by Vincent Scelta.
Camera-ready copy prepared by the author.
Printed and bound by Edwards Brothers, Inc., Ann Arbor, MI.
Printed in the United States of America.

9 8 7 6 5 4 3 2 1

ISBN 0-387-94181-9 Springer-Verlag New York Berlin Heidelberg
ISBN 3-540-94181-9 Springer-Verlag Berlin Heidelberg New York

Contents

CHAPTER 1

Distance and Angles

§ 1. Exercises

(Pages 5,6)

1. a. P $\longleftarrow$ Q

    b. P $\longleftarrow$ Q

    c. P $\underline{\hspace{3cm}}$ Q

    d. P $\longrightarrow$ Q

2. a.

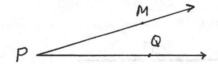

    b. The two rays, R_{PM} and R_{PQ}, form a whole line if M, P and Q
        are collinear and P lies between M and Q.

3. $d(P,Q) + d(Q,M) = d(P,M)$

4. $\overline{AB}$ is <u>not</u> parallel to $\overline{PQ}$ because L_{AB} is not parallel to L_{PQ}.

5. Line segments connecting any three points in the plane do not
    necessarily form a triangle since the three points may be collinear.

6.   If line L were parallel to line U there would be 2 lines parallel to line
      U passing through the point P. This contradicts PAR 2. Therefore
      line L is not parallel to line U and must intersect it.

7.   By the definition of parallelism, two lines, K and L, are parallel if K
      = L or K is not equal to L and does not intersect L. If P is on line L
      and K = L, then K and L are parallel lines.

Experiment 1-1
(Pages 7,8)

1.

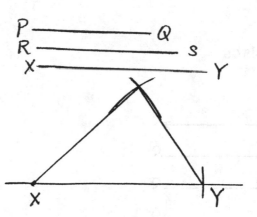

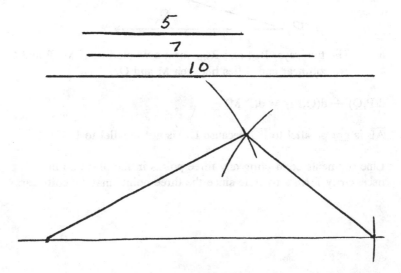

2.

3.

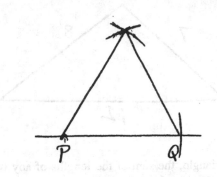

4.

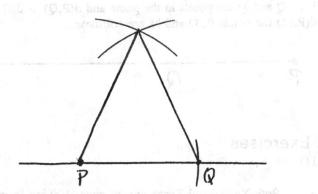

5.

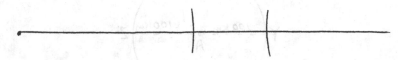

6. The arcs with radii 7 cm and 5 cm do not intersect since $5+7 < 15$.

7.   a.

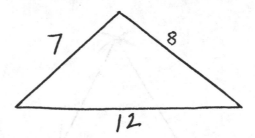

b.   For a triangle, the sum of the lengths of any two sides is greater than the length of the third side.

8.   You cannot construct a triangle with sides 5 cm, 10 cm and 15 cm since the segments would be collinear.

9.   If P, Q and M are points in the plane and d(P,Q) + d(Q,M) = d(P,M) the points P, Q and M are collinear.

§ 2. Exercises
(Page 11)

1.   a.   Both Ygleph and Zyzzx are, at most, 100 km from the antenna.

      b.   The messenger travelling from Ygleph to the antenna and then to Zyzzx would travel at most 200 km.

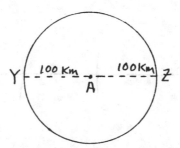

      c.   The maximum possible distance between Ygleph and Zyzzx is 200 km. This distance would occur if Ygleph and Zyzzx were at opposite ends of a diameter of a circle with radius 100 km. (see diagram in 1b.)

Proof:

$d(Y,Z) \leq d(Y,A) + d(A,Z)$ by the Triangle Inequality
$\leq 100 + 100$ since Y and Z receive the signal
≤ 200

2. $d(A,C) \leq d(A,B) + d(B,C)$ by the Triangle inequality
$d(A,C) \leq 265 + 286 = 551$ by the charts
Also $286 \leq d(A,C) + d(A,B)$ by the Triangle inequality
$\leq d(A,C) + 265$, since $d(A,B) = 265$.
Hence, $d(A,C) \geq 286-265=21$.
Therefore, $21 \leq d(A,C) \leq 551$

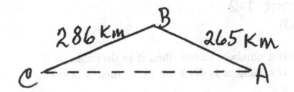

3.   (a)   yes
(b)   yes
(c)   no, $5 + 2 < 8$
(d)   yes
(e)   no, $1\ 1/2 + 3\ 1/2 = 5$, collinear
(f)   yes

4.   If two sides of a triangle are 12 cm and 20 cm, the third side must be larger than __8__ cm, and smaller than __32__ cm.

$20 - 12 < x < 12 + 20$, where x is the length of the third side

5. $d(P,Q) < r_1 + r_2$

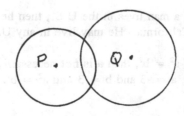

6.　　<u>X 1½ Z　　　　　　　Y</u>
　　　　　　　5

d(X,Z) + d(Z,Y) = d(X,Y), by SEG postulate.
Hence, 1 1/2 + d(Z,Y) = 5 and d(Z,Y) = 3 1/2.

7.　　Check your own work.

8.　　Since X and Y are contained in the disc, d(P,X) ≤ r and d(P,Y) ≤ r.
　　　d(X,Y) ≤ d(P,X) + d(P,Y) by the Triangle inequality
　　　Therefore, d(X,Y) ≤ r + r = 2r

Experiment 1-2
(Pages 12, 13)

1.　　.a.　　If a number is even, then it is divisible by 2.
　　　　　　If a number is divisible by 2, then it is even.

　　　b.　　If 6x = 18, then x = 3.
　　　　　　If x = 3, then 6x = 18.

　　　c.　　If a car is registered in California, then it has California license
　　　　　　plates.
　　　　　　If a car has California license plates, then it is registered in
　　　　　　California.

　　　d.　　If all of the angles of a triangle have equal measure, then the
　　　　　　triangle is equilateral.
　　　　　　If the triangle is equilateral, then all of its angles have equal
　　　　　　measure.

　　　e.　　If two distinct lines are parallel, then they do not intersect.
　　　　　　If two distinct lines do not intersect, then they are parallel.

2.　　a.　　<u>False</u>, If the square of a number is 9, then the number is not
　　　　　　necessarily 3 because $(-3)^2 = 9$ also.

　　　b.　　<u>False</u>, If a man lives in the U.S., then he does not necessarily
　　　　　　live in California. He may live in any U.S. state.

　　　c.　　<u>False</u>, If $a^2 = b^2$, then a is not necessarily equal to b. For
　　　　　　example, a = -3 and b = 3 and $a^2 = b^2$.

d.   <u>True</u>

e.   <u>False</u>, The statement <u>If the sum of integers x, y and z is 3y,</u>
      <u>then x, y and z are consecutive integers</u> is a false statement.
      For example,
      let x = 5, y = 10 and z = 15. Then x + y + z = 30 = 3y;
      but 5, 10 and 15 are not consecutive integers.

3.   The converse of a true "if-then" statement is <u>not</u> always true. Note #2
      (a, b, c and d).

      Make up five "if-then" statements and test them to determine whether
      they are true or not by testing the statement for any exceptions. If
      there are any exceptions, the "if-then" statement is false.

§ 3. Exercises
(Pages 22-25)

1.   a.   ∠ABC or ∠CBA or ∠B
      b.   ∠RPQ or ∠QPR
      c.   ∠YOX or ∠XOY
      d.   ∠BOC or ∠COB

2.   a.   x = 180 - 40 = 140°
      b.   x = 360 - 68 = 292°
      c.   x = 146 - 35 = 111°

3.   a.   b.

      c.   d.

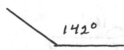   

e.

$270°$

4.   (1) 45°   (2) 60°   (3) 135°
     (4) 90°   (5) 20°   (6) 225°

5.   90/360 = 1/4   The length of arc is (1/4) (36) = 9.

6.   a.   45/360 = 1/8   The length of arc is
           (1/8)(36) = 9/2
     b.   180/360 = 1/2 The length of arc is
           (1/2)(36) = 18
     c.   60/360 = 1/6   The length of arc is
           (1/6)(36) = 6

     d.   (x/360)(36) = x /10

7.   (1/360)(40,000) = 40,000/360 = 1000/9 km

8.   a. Distance to the equator =
          (43/360)(40,000) = 172000/360 km = 43000/9 km
     b. (47/360)(40,000) = 47000/9 km to North Pole since 90°-43° = 47°

9.   If l is the latitude of your home city or town, take
     [(90 - l)/360](40,000) = distance to N.P.

10. All of the angles ∠ AXB have the same measure.

11. a. 30°   b. 220°   c. 110°

12. a.   b.

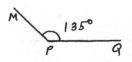

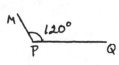

     c.   d.

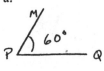

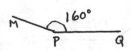

e.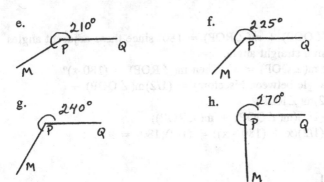

f. 225°

g. 240°

h. 270°

13.   Converse: If points Q, P and M lie on the same line, then m($\angle$ OPM) = 0°. It is false as shown by

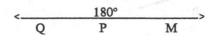

Experiment 1-3
(Pages 27-29)

1.   For example,

a.   m($\angle$ A) = m($\angle$ A') = 120°,
b.   m($\angle$ B) = m($\angle$ B') = 60°

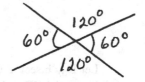

c.   Measure the angles as above.

d.   Conclusion: Opposite angles have the same measure.

2.   a.   Produce a figure similar to fig. 1.51

b.   The measure of the angle formed by the bisectors is 90°.

Exercise 1.
a.   m($\angle$ A) + m($\angle$ B) = 180° since they form a straight angle.
b.   For the same reason, m($\angle$ A') + m($\angle$ B) = 180°.
c.   m($\angle$ A) = m($\angle$ A')

Exercise 2.

a.   m($\angle$QOP) + m($\angle$ROP) = 180° since these adjacent angles form a straight angle.

b.   Let m($\angle$QOP) = x°. Then m($\angle$ROP) = (180-x)°
m(angle between bisectors) = (1/2)m($\angle$QOP) + (1/2)m($\angle$ROP)
     = (1/2)(m($\angle$QOP) + m($\angle$ROP))
     = (1/2)(x + (180 - x)) = (1/2)(180) = 90°.

Proof of 1.

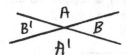

Because $\angle$A and $\angle$B form a straight angle, we know m($\angle$A) + m($\angle$B) = 180.   Similarly,
m($\angle$A') + m($\angle$B) = 180. Therefore,
m($\angle$A) + m($\angle$B) = m($\angle$A') + m($\angle$B).
m($\angle$A) = m($\angle$A') subtracting m($\angle$B) from each side.
When two lines intersect, opposite angles have the same measure.

Proof of 2

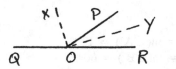

Let R_{OX} bisect $\angle$QOP and R_{OY} bisect $\angle$ROP.
Since $\angle$QOP and $\angle$ROP form a straight angle we know
m($\angle$QOP) = m($\angle$ROP) = 180. If we let m($\angle$QOP) = x and substitute in the first equation, we obtain
x + m($\angle$ROP) = 180. Thus
m($\angle$ROP) = 180 -x.
The angle between the bisectors, $\angle$XOY, has measure
m($\angle$XOY)= (1/2) m($\angle$QOP) + (1/2) m($\angle$ROP)
     = (1/2) (m($\angle$QOP) + m($\angle$ROP))
     = (1/2) (x + (180 - x))
     = (1/2) (180) = 90°.
Thus, we have that the bisectors of a pair of linear angles form an angle whose measure is 90°.

§ 4. Exercise
(Pages 33-36)

1.   A <u>postulate</u> is a statement accepted as true without proof.
    a.   Acceptable answers include all postulates given in the text. The
           Triangle Inequality Postulate is one example.
    b.   e.g. Given real numbers x, y and a, if $x=y$ then $x+a=y+a$.

2.   a. $m(\angle NOM) = m(\angle POQ) = 90 - 55 = 35°$
    b. $m(\angle SOQ) = m(\angle RON)$, Opposite angles

3. Given $m(\angle WPX) = 2m(\angle WPY) = 2x°$
   and $m(\angle WPY) = x°$. (see figure 1.57)
   $m(\angle YPV) = m(\angle WPX) = 2x°$, opposite angles
   $x + 2x = 180$, since $\angle WPX$ and $\angle WPY$ form a straight angle.
   Thus $3x = 180$ and $x = 60$. Therefore, $m(\angle YPV) = 120°$.

4.   Assume R_{OB} bisects $\angle AOC$ and R_{OC} bisects $\angle BOD$.
    Prove: $m(\angle AOB) = m(\angle COD)$(see fig.1.58)

    Let $m(\angle AOB) = x$, $m(\angle BOC) = y$ and $m(\angle COD = z$.
    Since R_{OB} bisects $\angle AOC$, $x = y$.
    Since R_{OC} bisects $\angle BOD$, $y = z$.
    Therefore $x = z$ and $m(\angle AOB) = m(\angle COD)$.

5. Assume lines $\overleftrightarrow{PV}$, $\overleftrightarrow{QT}$ and RS meet in point O and line $\overleftrightarrow{QT}$ bisects
   $\angle POR$. Prove: Line $\overleftrightarrow{QT}$ bisects $\angle SOV$. see fig.1.59)

    $m(\angle POQ) = m(\angle QOR)$ since R_{OT} bisects $\angle POR$
    $m(\angle POQ) = m(\angle TOV)$, opposite angles
    $m(\angle QOR) = m\angle SOT)$, opposite angles
    Therefore, $m(\angle TOV) = m(\angle SOT)$.
    Thus R_{OT} bisects $\angle SOV$ and line QT bisects $\angle SOV$.

6.   Assume: $d(P,Q) = d(R,S)$
    Prove: $d(P,R) = d(Q,S)$ (see fig. 1.60)

    Since $d(P,Q) = d(R,S)$ we can add $d(Q,R)$ to both sides of this
    equation and obtain
    $d(Q,R) = d(P,Q) + d(Q,R)$ by SEG postulate
           $= d(R,S) + d(Q,R)$ by assumption
           $= d(Q.S)$ by SEG postulate

7.   Assume points X and Y are contained in a disc with radius r around a
      point P. Prove: $d(X,Y) \leq 2r$.

      $d(X,Y) \leq d(P,X) + d(P,Y)$, by the Triangle Inequality
                $\leq r + r = 2r$ because X, Y are in the disc centered at P,
                      radius r

8.   Assume line L intersects lines K and U so that $\angle 1$ is supplementary to
      $\angle 2$. Prove: $m(\angle 3) = m(\angle 4)$ (see fig.1.61)

      $m(\angle 1)+m(\angle 2) = 180$, $\angle 1$ supplementary to $\angle 2$
      $m(\angle 1)+m(\angle 3) = 180$, $\angle 1$, $\angle 3$ are linear angles Therefore,
      $m(\angle 1)+m(\angle 2) = m(\angle 1)+m(\angle 3)$. Thus
            $(\angle 2) = m(\angle 3)$, by subtraction and
            $m(\angle 2) = m(\angle 4)$ since they are opposite angles.
            $m(\angle 3) = m(\angle 4)$ by substitution.

9.   Assume in triangle ABC, $m(\angle CAB) = m(\angle CBA)$ and in triangle
      ABD, $m(\angle DAB) = m(\angle DBA)$.
      Prove: $m(\angle CAD) = m(\angle CBD)$ (see fig.1.62)

      Since $\angle DAB$ and $\angle CAD$ are adjacent angles, we obtain
      $m(\angle DAB) + m(\angle CAD) = m(\angle CAB)$.
      $m(\angle CAD) = m(\angle CAB)-m(\angle DAB)$ by subtraction and
                   $= m(\angle CBA)-m(\angle DBA)$, assumption and substitution
                   $= m(\angle CBD)$ $\angle CBA$ and $\angle DBA$ are adjacent angles.
      Therefore, $m(\angle CAD) = m(\angle CBD)$

10.   Assume $m(\angle b) = m(\angle c)$
      Prove: $m(\angle a) = m(\angle d)$ (see fig. 1.63)

      $\angle a$ and $\angle b$ are opposite angles as are $\angle c$ and $\angle d$. By theorem 1.1
      $m(\angle a) = m(\angle b)$ and
      $m(\angle c) = m(\angle d)$. Then
      $m(\angle b) = m(\angle c)$ by assumption, and
      $m(\angle a) = m(\angle d)$ by substitution.

11.   Assume points P, B, C and Q are collinear, $m(\angle x) = m(\angle y)$,
      segment $\overline{BK}$ bisects $\angle ABC$ and segment $\overline{CK}$ bisects $\angle ACB$.
      Prove: $m(\angle KBC) = m(\angle KCB)$ (see fig. 1.64)

      Since $\angle PBA$ and $\angle ABC$ are linear angles we have
      $m(\angle x)+m(\angle ABC) = 180$
      $m(\angle ABC) = 180 - m(\angle x)$. Similarly, since

m($\angle$ ACB) = 180 - m($\angle$ y) since $\angle$ y and $\angle$ ACB are linear angles.
Segment $\overline{BK}$ bisects $\angle$ ABC so
m($\angle$ KCB) = 1/2 m($\angle$ ABC) and
            = 1/2(180 - m($\angle$ x)). Thus
            = 1/2(180 - m($\angle$ y), by assumption.
Also, segment $\overline{CK}$ bisects $\angle$ ACB and
m($\angle$ KCB) = 1/2 m($\angle$ ACB) Thus
            = 1/2(180 - m($\angle$ y) by substitution.
m($\angle$ KBC) = m($\angle$ KCB) since both equal 1/2(180 - m($\angle$ y).

§ 5. Exercise
(Pages 41-43)

1. Since K $\perp$ V and L $\perp$ V we have K parallel to L by theorem 1.2.
   Two lines perpendicular to the same line in a plane are parallel.

2. Given that K$\perp$L$_1$ and L$_1$ is parallel to L$_2$ we have K$\perp$L$_2$ by PERP 2.
   Given two parallel lines, L$_1$ and L$_2$, if K$\perp$L$_1$, then K$\perp$L$_2$.

3. Assume $\overrightarrow{PR} \perp \overrightarrow{PT}$ and $\overrightarrow{PQ} \perp \overrightarrow{PS}$
   Prove: m($\angle$ a) = m($\angle$ b) (see fig. 1.76)

   By assumption, $\overrightarrow{PR} \perp \overrightarrow{PT}$ thus $\angle$ TRP is a right angle and
   m($\angle$ b)+m($\angle$ x)=90. Since $\overrightarrow{PQ} \perp \overrightarrow{PS}$ we also have
   m($\angle$ a)+m($\angle$ x)=90. Since both sums equal 90 we obtain
   m($\angle$ a)+m($\angle$ x)=m($\angle$ b)+m($\angle$ x). By subtraction m($\angle$ a)=m($\angle$ b).

4. For each example, the assumption should lead to a contradiction of
   some accepted statement (as in the example).

5. Assume: ABCD is a parallelogram with $\angle$ A a right angle (see fig.
   1.77) Prove: $\angle$ B, $\angle$ C and $\angle$ D are right angles.

   Since $\angle$ A is a right angle, $\overline{AB} \perp \overline{AD}$. Since ABCD is a parallelogram
   $\overline{AD}$ is parallel to $\overline{AB}$. By Perp 2 we conclude that $\overline{AB} \perp \overline{BC}$ and $\angle$ B
   is a right angle. Also, $\overline{AB}$ is parallel to $\overline{CD}$ and BC$\perp$AB so Perp 2
   implies that $\overline{BC} \perp \overline{CD}$ and $\angle$ C is a right angle. Finally, since
   $\overline{CD} \perp \overline{BC}$ and $\overline{BC}$ is parallel to $\overline{AD}$ we obtain $\overline{CD} \perp \overline{AD}$ and $\angle$ D is a
   right angle by Perp 2. Therefore, $\angle$ B, $\angle$ C and $\angle$ D are right angles
   and ABCD is a rectangle.

6. a) 90
   b) 90 - 50 = 40 (see fig. 1.78)
   c) yes
   d) 180
   e) 90 - 23 = 67
   f) $\angle 1$ and $\angle 3$
   g) $m(\angle 2) = 90 - 32 = 58$
      $m(\angle TOB) = 58 + 90 = 148$
   h) If $m(\angle 1) + m(\angle 4) = 90$ by (d) we know $m(\angle 2) + m(\angle 3) =$ 90 and R_{OT} is perpendicular to R_{OS}.

7. Using Post Perp 1 we can conclude that L_1 and L_2 are the same line.

8. In quadrilateral PBQC assume $\angle PBQ$ and $\angle PCQ$ are right angles and $m(\angle x) = m(\angle y)$. Prove: $m(\angle ABQ) = m(\angle DCQ)$ (see fig. 1.79)

   $m(\angle x) + m(\angle CBQ) = 90$ given the $\angle PBQ$ is a right angle.
   $m(\angle y) + m(\angle BCQ) = 90$ since $\angle PCQ$ is a right angle. By addition
   $m(\angle x) + m(\angle CBQ) = m(\angle x) + m(BCQ)$ since $m(\angle x) = m(\angle y)$.
   $m(\angle CBQ) = m(\angle BCQ)$ by subtraction. Then
   $m(\angle CBQ) + m(\angle ABQ) = 180$ $\angle CBQ$ and $\angle ABQ$ are linear angles.
   $m(\angle BCQ) + m(\angle DCQ) = 180$ again by linear angles.
   $m(\angle CBQ) + m(\angle ABQ) = m(\angle CBQ) + m(\angle DCQ)$ they equal 180.
   $m(\angle ABQ) = m(\angle DCQ)$ by subtraction.

9. Assume L_1 is parallel to L_2 and L_2 is parallel to L_3.
   Prove: L_1 is parallel to L_3

   Further assume that L_1 is not parallel to L_3. By PAR 1, L_1 meets L_3 in a point, P. By PAR 2, given line L_2 and point P there is one and only one line passing through P parallel to L_2. However, L_1 and L_3 both pass through point P and are parallel to L_2. This contradicts PAR 2 so we conclude that the above assumption is false and L_1 is parallel to L_3.

§ 4. Experiment
(Pages 45-46)

1.

$m(\angle a) + m(\angle b) = 54 + 36 = 90$

2.   m($\angle$ a) + m($\angle$ b) = 38 + 52 = 90
      m($\angle$ a) + m($\angle$ b) = 77 + 13 = 90

3.   No, Two right triangles with corresponding legs of equal measures
      have the same shape.

4.   Two such right triangles form a rectangle.

5.   The m($\angle$ a)+m($\angle$ b)+m($\angle$ c)=180

6.   Constructing a perpendicular at a point on a line is like bisecting a
      straight angle.

§ 6. Exercise
(Pages 54-60)

1.   a.   m($\angle$ C) = 90 - 34 = 56   b. m($\angle$ C) = 30
      c.   m($\angle$ C) = 60   d. m($\angle$ C) = 45

2.   a.   m($\angle$ C)=180 - (47+110) = 23
      b.   m($\angle$ C)=30   c. m($\angle$ C)=50   d. m($\angle$ C)=36

3.   a.   m($\angle$ x)=180 - 50=130   b. m($\angle$ x)=80
      c.   m($\angle$ x)=90 - 64=26   d. m($\angle$ x)=180-60=120

4.   a.   m($\angle$ x)=180-130=50   b. m($\angle$ x)=180-18=162
      c.   m($\angle$ x)=360-(100+95+60)=105
      d.   m($\angle$ x)=180-83=97

5.
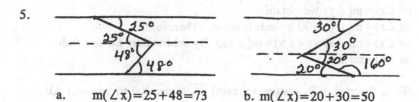

      a.   m($\angle$ x)=25+48=73   b. m($\angle$ x)=20+30=50

6.   Let m($\angle$ A)=x, m($\angle$ B)=2x and m($\angle$ C)=3x. Then
      x+2x+3x=180 and 6x=180. Thus x=30 and the three angles have
      measure 30, 60 and 90.

7.   (see figure 1.106) Since $\overline{AP}$ bisects $\angle A$ let $m(\angle CAP)=m(\angle PAB)=x$.
      Similarly, let $m(\angle CBP)=m(\angle PBA)=y$. Then
      $2x+2y+70=180$ and
      $2x+2y=110$ or $x+y=55$.
      In $\triangle APB$, $x+y+m(\angle P)=180$. Combining these expressions
      $55+m(\angle P)=180$. Thus,
      $m(\angle P)=180-55=125$.

8.   Since a square and a rectangle can be divided up into two triangles, the
      sum of the angles of a square and of a rectangle is 360°.

9.   Let SWAT be any four-sided figure. (see fig. 1.107)
      Prove: The sum of angles of SWAT is 360°.

      Connect the vertices W and T with a line segment. Note that this
      forms two triangles, STW and TWA. $\angle$ STW and $\angle$ WTA form
      $\angle$ STA and $\angle$ SWT and $\angle$ TWA form $\angle$ SWA. For each triangle the
      sum of angles is 180° so the total angle measure for SWAT is 2 x 180°
      = 360°.

10.   Suppose triangle ABC has two obtuse angles, $\angle A$ and $\angle B$. $m(\angle A)$
      + $m(\angle B)$ > 180° contradicting the theorem that the sum of angles in
      a triangle equals 180°. Therefore, no triangle can contain two obtuse
      angles.

11.   Yes, for example, the 30°, 60°, 90° triangle contains no obtuse angle.

12.   Assume that line M is perpendicular to line L, $m(\angle x)=m(\angle y)$ and L
      is parallel to K. Prove: $m(\angle a)=m(\angle b)$. (see fig. 1.108)

      Since L is parallel to K and M is perpendicular to L, by PERP 2 we
      know that M is perpendicular to K. Thus
      $m(\angle x)+m(\angle a)=90$ and $m(\angle y)+m(\angle b)=90$. Because
      $m(\angle x)=m(\angle y)$ we obtain
      $m(\angle x)+m(\angle b)=90$ by substitution. Therefore,
      $m(\angle x)+m(\angle a)=m(\angle x)+m(\angle b)$. By subtraction we conclude
      $m(\angle a)=m(\angle b)$.

13.   Given figure 1.109, assume segment $\overline{CL}$ is parallel to segment $\overline{AB}$.
      Prove: $m(\angle s)=m(\angle p)+m(\angle q)$

      $\angle r$ and $\angle BCL$ are alternate angles for the parallel lines $\overleftrightarrow{AB}$ and $\overleftrightarrow{CL}$.
      Thus, $m(\angle r)=m(\angle BCL)$. Also, $m(\angle BCL)+m(\angle s)=180$ since they
      form a linear pair of angles. By substitution we obtain

m($\angle$ r)+m($\angle$ s)=180. In the triangle ABC, m($\angle$ r)+m($\angle$ p)+m($\angle$ q)=180. Therefore, m($\angle$ r)+m($\angle$ s)=m($\angle$ r)+m($\angle$ p)+m($\angle$ q). By subtraction we conclude m($\angle$ s)=m($\angle$ p)+m($\angle$ q).

14. Let PMQ be a right triangle with $\angle$ M the right angle and assume segment $\overline{MH}$ is perpendicular to segment $\overline{PQ}$. (see fig. 1.110) Prove: The 3 angles of $\triangle$PHM have the same measure as the 3 angles of $\triangle$MHQ.

m($\angle$ PHM)=m(MHQ) because they are both right angles. Since $\angle$ Q belongs to both triangles, PMQ and MHQ we note that m($\angle$ Q)+m($\angle$ HMQ)+90 =180 and m($\angle$ Q)+m($\angle$ P)+90=180. Therefore, by subtraction we obtain m($\angle$ HMQ)=m($\angle$ P). Finally, since two pairs of angles of triangles PMH and MHQ have the same measure the remaining pair of angles will also have the same measure. Therefore, the three angles of $\triangle$PHM have the same measures as the three angles of $\triangle$MHQ.

15. Let TALK be a parallelogram.
Prove: m($\angle$ K)=m($\angle$ A) and m($\angle$ T)=m($\angle$ L)

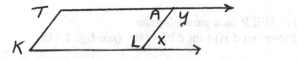

Construct the rays R_{TA} and R_{KL}. Name the new angle adjacent to $\angle$ A, $\angle$ y and the other new angle, $\angle$ x. $\angle$ T and $\angle$ y are parallel angles so m($\angle$ T)=m($\angle$ y). And
m($\angle$ L)=m($\angle$ y) because they are alternate angles. Thus,
m($\angle$ T)=m($\angle$ L). In a similar way
m($\angle$ K)=m($\angle$ x) and
m($\angle$ A)=m($\angle$ x) so
m($\angle$ K)=m($\angle$ A).

16. Theorem 1-7

17. Let ABC be an arbitrary triangle with side $\overline{AC}$ extended in the C direction.
Prove: m($\angle$ 1)=m($\angle$ A)+m($\angle$ B) (see fig. 1.112)

Note that $\angle$ 1 and $\angle$ C form a linear pair so m($\angle$ 1)+m($\angle$ C)=180. In $\triangle$ABC we have m($\angle$ A)+m($\angle$ B)+m($\angle$ C)=180. Thus, m($\angle$ 1)+m($\angle$ C)=m($\angle$ A)+m($\angle$ B)+m($\angle$ C). By subtraction we obtain our conclusion since m($\angle$ 1)=m($\angle$ A)+m($\angle$ B).

18.   Assume for $\angle A$ two perpendiculars are dropped from point Q (within the angle) to each of the rays intersecting the rays at points M and N. (see fig. 1.113) Prove: $\angle A$ and $\angle C$ are supplementary.

Since $\overline{QM} \perp \overline{AM}$ and $\overline{QN} \perp \overline{AN}$ the angles $\angle AMQ$ and $\angle ANQ$ are right angles. Thus $m(\angle AMQ) + m(\angle ANQ) = 180$. The quadrilateral's four angles have measures adding up to $360°$ so $m(\angle A) + m(\angle C) = 180$ by subtraction and we can conclude that $\angle A$ and $\angle C$ are supplementary.

19.   In figure 1.114, segment $\overline{QN}$ is perpendicular to R_{PS} and segment $\overline{QM}$ is perpendicular to R_{PT}.
      Prove: $m(\angle P) = m(\angle Q)$

$m(\angle QNX) = m(\angle PMX)$ since they are right angles.
$m(\angle QXN) = m(\angle PXM)$ since they are opposite angles.
$m(\angle QNX) + m(\angle QXN) + m(\angle Q) = 180$ and
$m(\angle PMX) + m(\angle PXM) + m(\angle P) = 180$. These sums are equal so by subtraction of equal terms from the both sides of that equation, thus we conclude $m(\angle P) = m(\Delta Q)$.

20.   Let HELP be a parallelogram.
      Prove: $m(\angle H) + m(\angle P) = 180$ (see fig. 1.115)

Opposite angles of a parallelogram have equal measure. Therefore, $m(\angle E) = m(\angle P)$ and $m(\angle H) = m(\angle L)$. In a parallelogram $m(\angle H) + m(\angle E) + m(\angle L) + m(\angle P) = 360$. By substitution, $2(m(\angle H) + m(\angle P)) = 360$ and $m(\angle H) + m(\angle P) = 180$. Thus $\angle H$ and $\angle P$ are supplementary angles.

21.   Let two lines L_1 and L_2 be cut by a third line, K, so that $m(\angle A) = m(\angle B)$.
      Prove: L_1 and L_2 are parallel. (see fig. 1.116)

Name the angle opposite $\angle A$, $\angle C$.
$m(\angle A) = m(\angle C)$ because they are opposite angles.
$m(\angle A) = m(\angle B)$ by assumption. And
$m(\angle B) = m(\angle C)$ by substitution. By theorem 1-6 and the equality of the measures of these parallel angles, L_1 is parallel to L_2.

22.   In figure 1.117, L_1 is parallel to L_2 and $\overline{XT} \perp L_2$ and $\overline{AB} \perp L_1$.
      Prove: $m(\angle A) = m(\angle X)$

Construct the ray R_{XY}. Since $R_{XY} \perp L_2$ we conclude by PERP 2 that $R_{XY} \perp L_1$. $\overline{AB} \perp L_1$ so by theorem 1-2 we conclude that R_{XY} is parallel to AB. $\angle A$ and $\angle X$ are alternate angles for these parallel lines so $m(\angle A) = m(\angle X)$.

23.   In the right triangle in figure 1.118, $\overline{CN} \perp \overline{AB}$, the hypotenuse. Prove: $m(\angle NCB) = m(\angle A)$

In right triangle, ABC, $\angle ACB$ is the right angle. Thus $m(\angle ACB) = m(\angle CNB)$ since $CN \perp AB$. $\angle B$ is an angle of $\triangle ABC$ and $\triangle NCB$. Consequently, $m(\angle ACB) + m(\angle B) + m(\angle A) = 180$ and $m(\angle CNB) + m(\angle B) + m(\angle NCB) = 180$ Then $m(\angle A) = m(\angle NCB)$ by subtraction of equals from equals.

ADDITIONAL EXERCISES FOR CHAPTER 1

1.   a)   56° b) 129° c) 330°

2.   The bisectors of the angles of a triangle are concurrent.

3.   Circle (a) and (b).

4.   (a)   P, Q, and M are on the ray R_{PM} and are collinear.
      (b)   P, Q, and M are collinear.

5.   Points X and Y do not have to be on the disc at all or one point could be on the disc and the other off the disc. If both points were on the disc, they would be endpoints of a diameter.

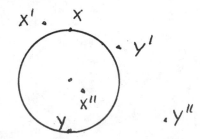

6.   (a)   180° - 63° = 117° (b) 73°
      (c)   93° (d) 137°

7.   Assume L_1 is parallel to L_2.
      Prove: $\angle A$ is supplementary to $\angle B$. (see fig. 1.121)

Let $\angle C$ be the angle adjacent to $\angle A$. Since $\angle A$ and $\angle C$ form a linear pair $m(\angle A) + m(\angle C) = 180$. $\angle B$ and $\angle C$ are parallel angles so $m(\angle B) = m(\angle C)$. By substitution we obtain, $m(\angle A) + m(\angle B) = 180$. Therefore $\angle A$ is supplementary to $\angle B$.

8.   Suppose the right triangle ABC has its right angle at C and an obtuse
     angle at A. m($\angle$ A)>90 and m($\angle$ C)=90. Thus
     m($\angle$ A)+m($\angle$ C)>180 which is impossible. Therefore, a right
     triangle cannot have an obtuse angle.

9.   Assuming that you can draw a perpendicular to any two lines is
     impossible if they are not parallel. Thus in this "proof" there is an
     unsupported assumption that L_1 is parallel to L_2.

CHAPTER 2

Coordinates

§ 1. Exercises
(Pages 70-71)

1.

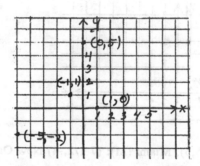

2.

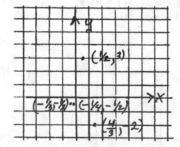

3.   If (x,y) are the coordinates of a point in the second quadrant, x < 0 and y > 0.

4.   If (x,y) are the coordinates of a point in the third quadrant, x < 0 and y < 0.

5.

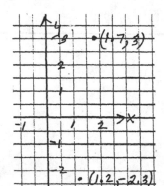

6.

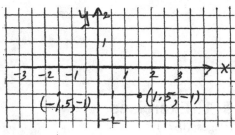

7.

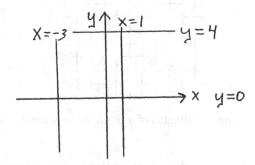

8.   Seattle (S): (-5,5), Miami (M): (7 1/2,-4) New Orleans (N): (3, -3)

9.   (a)   y-axis, (b) x-axis, (c) x-axis,
     (d)   both x and y axes

10.

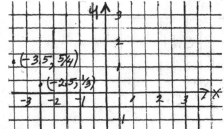

11. (a)   The line x = 3 is _parallel_ to the line x = -3.
     (b)   The line y = -7 is _perpendicular_ to the y-axis.
     (c)   The line y = -7 is _parallel_ to the x-axis.
     (d)   The line y = 4 is _perpendicular_ to the line x = 7.

12. The line x = 0 is equal to the y-axis.
    The line y = 0 is equal to the x-axis.

13. (a) ▵ABC is a right triangle

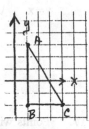

(b) ▵ABC is a right triangle

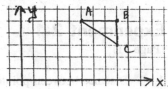

(c) ▵ABC is <u>not</u> a right triangle

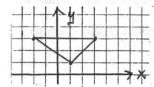

14. (a) ▵XYZ is an isosceles triangle

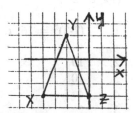

(b) ▵XYZ is <u>not</u> an isosceles triangle.

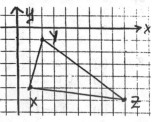

(c) ▵XYZ is an isosceles triangle

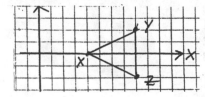

15. The fourth coordinate pair to form the parallelogram could be (11,5), (-5,5) or (5,-5).

16.   The set of points with x and y coordinates that are equal form a line y=x.

17.   Points with the y-coordinate twice the x-coordinate form a line y = 2x.

Experiment 2-1
(Page 71)

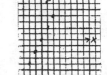

1.   Distance between points 3 and 5 equals 2.

2.   Distance between points -2 and 4 equals 6.

3.   Distance between points -3 and -1 equals 2.

§ 2. Exercises
(Page 72)

1. $(7-3)^2 = 16$, d = 4, the square root of 16.

2.   10
3.   10
4.   4
5.   2
6.   3
7.   3

8.   (a)   4 or 6
       (b)   0 or 10
       (c)   -1 or 11
       (d)   -2 or 12

9.   (a)   -4 or 0
       (b)   -5 or 1
       (c)   -6 or 2
       (d)   -7 or 3

10.   h-8 or h+8

§ 3. Exercise

(Pages 79, 80)

1.

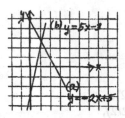

2.

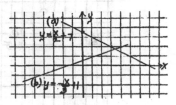

3.

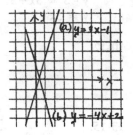

4.   (a)   slope = (-7 - 1)/(2 - (-1)) =
$$-8/3(y - 1) = (-8/3)(x - (-1))$$
$$y - 1 = (-8/3)x - (8/3)$$
$$y = (-8/3)x - (5/3)$$

     (b)   slope = ((1/2) - (-1))/(3-4) = (3/2)/(-1) = -3/2
$$(y-(-1)) = (-3/2)(x-4)$$
$$y+1 = (-3/2)x + 6$$
$$y = (-3/2)x + 5$$

5.   (a)   slope = (1 - (-1))/($\sqrt{2}$ - $\sqrt{2}$) = 2/0 therefore a vertical line
$x = \sqrt{2}$.

     (b)   slope = (4 - (-5))/($\sqrt{3}$ - (-3)) = 9/($\sqrt{3}$+3)
$$(y - (-5)) = (9/(\sqrt{3}+3))(x - (-3))$$
$$y+5 = (9/(\sqrt{3}+3))x + 27/(\sqrt{3}+3)$$
$$y = (9/(\sqrt{3}+3))x + (27/(\sqrt{3}+3)) - 5$$
$$y = (9/(\sqrt{3}+3))x + (27 - 5(\sqrt{3}+3))/(\sqrt{3}+3)$$

6. $y - 1 = 4(x - 1)$,
    $y - 1 = 4x - 4$,
    $y = 4x - 3$

7. $y - 1 = -2(x - (1/2))$,
    $y - 1 = -2x + 1$,
    $y = -2x + 2$

8. $y - 3 = (-1/2)(x - \sqrt{2})$,
    $y - 3 = (-1/2)x + (\sqrt{2}/2)$,
    $y = (-1/2)x + (6 + \sqrt{2})/2$

9. $y - 5 = 3(x - (-1))$,
    $y - 5 = 3x + 3$,
    $y = 3x + 8$

10-12.

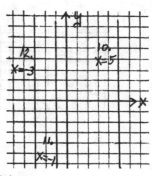

13-15.

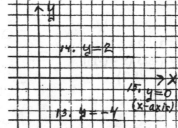

16. slope = -1/4

17. slope = -4

18. slope = 0

19. slope = $1/(3 - \sqrt{3})$

20. $y - 3 = (2/(\sqrt{2} - \pi))(x - \sqrt{2})$

21. $y - 2 = ((\pi - 2)/(1 - \sqrt{2}))(x - \sqrt{2})$

22. $y - 2 = (-3/(\sqrt{2} + 1))(x - (-1))$

23. $y + 3 = ((\sqrt{2} - (-3))/1)(x + 2)$

24.

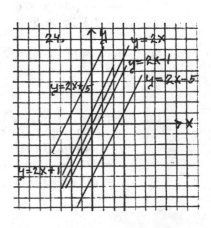

CHAPTER 3

Area and the Pythagoras Theorem

§ 1. Exercises
(Pages 91-94)

1. (a) $A = \frac{1}{2}(12)(5) = 30 \text{ cm}^2$
    (b) $A = \frac{1}{2}(3)(7) = 21/2 \text{ m}^2$
    (c) $A = \frac{1}{2}(2x)(x) = x^2$

2. Rectangle area $= 10 \times 40 = 400$, the side of the square with the same area is 20.

3. $A = 36 = \frac{1}{2}(b)(\frac{1}{2}b) = b^2/4$, $b^2 = (4)(36)$. Therefore, $b = 12$ m.

4. $A = w(5w) = 1440$, $5w^2 = 1440$, $w^2 = 288$ and $w = 12\sqrt{2}$. Thus the width is $12\sqrt{2}$ cm and the length is $60\sqrt{2}$ cm.

5. (a) $A = 4(5)(4) + 5^2 = 80 + 25 = 105 \text{ cm}^2$
    (b) $A = 4lh + l^2 = l(l+4h)$ (see fig. 3.17)

6. $A = (22 \times 17) - (20 \times 15) = 374 - 300 = 74 \text{ m}^2$

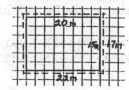

7. Let $w = 3x$ and $l = 4x$ thus $w:l = 3:4$.
    $A = (3x)(4x) = 300$, $12x^2 = 300$, $x^2 = 25$,
    $x = 5$ cm. Therefore, $w = 15$ cm and $l = 20$ cm.

8. $A = \frac{1}{2}(b)(12) = 36$, $6b = 36$, $b = 6$ cm.

9. $A = \frac{1}{2}(l)(l) = 40$, $l^2 = 80$, $l = 4\sqrt{5}$.

10.   Let a, b be the lengths of legs $\overline{AC}$ and $\overline{BC}$. Let 2a and 2b be the lengths of legs $\overline{XZ}$ and $\overline{ZY}$. Area of $\triangle ABC = \frac{1}{2}ab$, Area of $\triangle XYZ = \frac{1}{2}(2a)(2b) = 2ab$. Therefore, $A_{ABC}:A_{XYZ} = \frac{1}{2}ab:2ab = 1:4$

11. $1:n^2$ (Note the pattern in 10.)

12. $A = 8^2 = 64$ sq. units. $A = 3(64) = 192$.
      $A = 192 = s^2$, $s = 8\sqrt{3}$ units.

13.   Since the floor of the pool is 12 tiles by 18 tiles, the number of tiles needed is 216 tiles. At a cost of \$35 per tile, the total cost would be \$35(216) = \$7560.

14.   Area of rectangle = 12 x 5 = 60 (see fig. 3.19). Area of lower triangle = $\frac{1}{2}(5)(12)=30$.
      Area of small rectangle = 1 x 4 = 4. The last small rectangle has area $A = 2$ x $3 = 6$.
      Therefore the shaded area = 60-(30+4+6)=20.

15.   Area of $\triangle ANB = \frac{1}{2}(b+x)(h)=\frac{1}{2}bh+\frac{1}{2}xh$. Area of $\triangle ANC = \frac{1}{2}(xh)$.
      Therefore area of $\triangle ABC = \frac{1}{2}bh+\frac{1}{2}xh-\frac{1}{2}xh = \frac{1}{2}bh$.

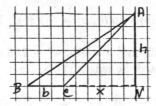

16. $A=\frac{1}{2}h(b+B)=\frac{1}{2}(7)(8+B)=77$, $(7)(8+B)=154$, $8+B=22$ thus $B=14$.

17.   Area of side wall = $(4)(5)+\frac{1}{2}(26)(1+5) =$
      20 + 78 = 98 sq. units.   (See figure 3.20)

18. $A=\frac{1}{2}(25)(30+40)+\frac{1}{2}(25)(40+55)+\frac{1}{2}(25)(55+75)$
      $A=\frac{1}{2}(25)(30+40+40+55+55+75)=\frac{1}{2}(25)(295) =$
      3687.5 m^2.

19.   Let N be the midpoint of side $\overline{QM}$. Thus, if $d(Q,N)=b$, $d(N,M)=b$.
      $\triangle QPN$ has the same height as $\triangle PNM$. Since the bases of $\triangle QPN$ and $\triangle NPM$ have equal measure and the heights are equal, their areas are equal and equal $\frac{1}{2}bh$.

20. (a) In figure 3.23, let d(Q,N)=b and d(N,M)=2b. Let h be the height of each of the triangles. Area of ΔQPN=½bh. Area of ΔPNM=½(2b)(h)=bh. Therefore ΔPNM has twice the area of ΔQPN.

(b) If N' is 2/3 of the distance from Q to M, d(N',M)=b. Area of ΔPMN'=½bh and is equal to the area of ΔQPN.

(c) Area of ΔPNN'=½bh=area of ΔPQN.

21. Using the pattern from question 20, area of ΔPNM is three times the area of ΔPQN. If the distance is one-fifth, ΔPNM has a base 4 times as long as the base of ΔPQN so its area is four times the area of ΔPQN.

22. In the figure, the diagonals are perpendicular. Let d(B,C)=b. $\overline{AM}$ is the altitude of ΔABC and $\overline{MD}$ is the altitude of ΔBCD.
The area of the quadrilateral
= ½(b)|AM|+½(b)|MD| and by the distributive property
= ½(b)(|AM|+|MD|) By joining adjacent segments we obtain
= ½|BC||AD| Therefore the area of the quadrilateral
= ½(product of the diagonals).

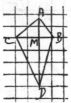

Experiment 3-2
(Page 94)

1.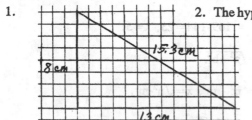
2. The hypotenuse is approximately 15.3 cm.

3.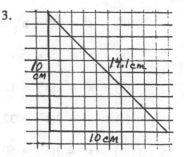

4. No, it is false. In Problem 1, 8+13>15.3.
In problem 3, 10+10>14.1.

5. $a^2 + b^2 = c^2$.

6.

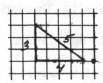

7.

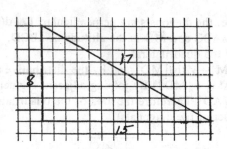

§ 2. Exercises
(Pages 101-108)

1. (a) $5^2 + x^2 = 36$, $x^2 = 11$, $x=\sqrt{11}$

    (b) $x^2 + 12^2 = 13^2$, $x^2 = 169-144=25$, $x=5$.

    (c) $a^2 + b^2 = x^2$, $x = \sqrt{a^2+b^2}$.

    (d) $x^2 + m^2 = n^2$, $x^2 = n^2 - m^2$, $x=\sqrt{n^2-m^2}$.

2. (a) $x^2 = 2^2 + 2^2 = 8$, $x = 2\sqrt{2}$.

    (b) $x^2 = 2(3^2) = 18$, $x = 3\sqrt{2}$.

    (c) $x^2 = 2(4^2) = 32$, $x = 4\sqrt{2}$.

    (d) $x^2 = 2(5^2) = 50$, $x = 5\sqrt{2}$.

    (e) $x^2 = 2(r^2)$, $x = r\sqrt{2}$.

3. (a) $d^2 = 1^2 + 2^2 = 5$, $d = \sqrt{5}$.

    (b) $d^2 = 3^2 + 5^2 = 34$, $d = \sqrt{34}$.

    (c) $d^2 = 4^2 + 7^2 = 65$, $d = \sqrt{65}$.

    (d) $d^2 = r^2 + (2r)^2 = 5r^2$, $d = r\sqrt{5}$.

    (e) $d^2 = (3r)^2 + (5r)^2 = 34r^2$, $d = r\sqrt{34}$.

    (f) $d^2 = (4r)^2 + (7r)^2 = 65r^2$, $d = r\sqrt{65}$.

4. Since the cube has the same length sides the formula for the diagonal   is
    (diagonal of base)^2 + (height of cube)^2 = $(s^2 + s^2)$ +s^2 which simplifies
    to $d^2 = 3s^2$.
    (a) $d^2 = 3(1)^2 = 3$, $d = \sqrt{3}$.
    (b) $d^2 = 3(2)^2 = 12$, $d = 2\sqrt{3}$.
    (c) $d^2 = 3(3)^2 = 27$, $d = 3\sqrt{3}$.
    (d) $d^2 = 3(4)^2 = 48$, $d = 4\sqrt{3}$.
    (e) $d^2 = 3r^2$, $d = r\sqrt{3}$.

5. In the drawing of the box, note x is the length of the diagonal of the base
    rectangle and d is the length of the diagonal of the box.

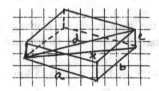

$x^2 = a^2 + b^2$
$d^2 = x^2 + c^2$ Thus,
$d^2 = a^2 + b^2 + c^2$.

    (a) $d^2 = 3^2 + 4^2 + 5^2 = 9+16+25=50$, $d=5\sqrt{2}$.
    (b) $d^2 = 1+4+16 = 21$, $d=\sqrt{21}$
    (c) $d^2 = 1+9+16 = 26$, $d=\sqrt{26}$.
    (d) $d^2 = a^2 + b^2 + c^2$, $d=\sqrt{a^2+b^2+c^2}$.
    (e) $d^2 = (ra)^2+(rb)^2+(rc)^2$, $d=r\sqrt{a^2+b^2+c^2}$.

6. In figure 3.34, note that the tower and ground form a right angle and the
    wire is the hypotenuse of the right triangle. Therefore,
    $w^2 = 25^2 + 30^2 = 625+900 = 1525 = (25)(61)$
    $w = 5\sqrt{61}$ m. That is approximately 39 m.

7.   Given that the baseball diamond is a square with sides 90 feet in length, the distance from the home plate to second base is the diagonal of the square.

$d^2 = 90^2 + 90^2 = 2(90)^2$, $d = 90\sqrt{2}$ feet.   This distance is approximately 127.28 feet.

8.   For this square the area = 144 sq. cm. so
$s^2 = 144$ and $s = \sqrt{144} = 12$. Let the diagonal be of length d.
$d^2 = 2(12)^2$ thus $d = 12\sqrt{2}$ cm.

9.   For this right triangle, let the shorter leg have length x. The other leg is then of length 2x. Since the area equals 72, $\frac{1}{2}x(2x)=72$, $x^2 = 72$, $x=6\sqrt{2}$ and $2x= 12\sqrt{2}$.
Let the hypotenuse have length h.
$h^2 = (6\sqrt{2})^2 + (12\sqrt{2})^2 = 72+288=360$, $h=6\sqrt{10}$.

10.   (a) $|PA_1|^2 = 1^2 + 1^2 = 2$, $|PA_1| = \sqrt{2}$.

      (b) $|PA_2|^2 = (\sqrt{2})^2 + 1^2 = 3$, $|PA_2| = \sqrt{3}$.

      (c) $|PA_3|^2 = (\sqrt{3})^2 + 1^2 = 4$, $|PA_3| = \sqrt{4} = 2$.

      (d) $|PA_n|^2 = (\sqrt{n})^2 + 1^2 =n+1$, $|PA_n| = \sqrt{(n+1)}$.

11.   The distance of the ship from its starting point is the length of the diagonal of a rectangle with sides 10 km and 5 km. See figure 3.36.

      (a) $d^2 = 10^2 + 5^2 = 125$, $d = 5\sqrt{5}$ km.

      (b)   Actual distance from the North Pole is 10 km. The eastward part does not take the ship further from the North Pole.

12.   (a) $h^2 = 3^2 + 4^2 = 25$, $h = 5$.

      (b) $h^2 = 6^2 + 8^2 = 100$, $h = 10$.

      (c) $h = 15$.

      (d) $h = 5c$.

      (e) $h^2 = (3c)^2 + (4c)^2 = c^2(3^2+4^2) = c^2(25)$. $h = 5c$.

      (f)   (i) $a = (100)(3)$, $b = (100)(4)$ so $c= (100)(5) = 500$
            (ii) 30   (iii) 45

13.   The authors suggest letting x=a/c and y=b/c so that $x^2 + y^2 = 1$. Furthermore, they suggest the formulas $x=(1-t^2)/(1+t^2)$ and $y=(2t)/(1+t^2)$. Does $x^2 + y^2 = 1$?

        Choose t=2: x=(-3)/5, y=4/5, a=3, b=4, c=5
        Choose t=7: x=(-48)/50, y=14/50 so a=14, b=48 and c=50 (Better to divide by 2) a=7, b=24, c=25.

14.   (a)   Divide 15 and 36 by 3 revealing 5, 12, 13. Thus x =3(13) = 39.

        (b)   (3, 4, 5) so x = 5 m.

        (c)   Divide 108 and 144 by 36 obtaining 3, 4, 5. Thus x = 36(5) = 180.

15.   Area of the right triangle equals 27 and its legs could have lengths 2x and 3x and be in the ratio 2:3. Thus, ½(2x)(3x)=27, $3x^2=27$. $x^2=9$ so x=3. The legs have lengths 6 and 9. $h^2 = 6^2 + 9^2 = 36+81=117$. $h = \sqrt{117} = 3\sqrt{13}$.

16.   Let this right triangle have legs with lengths 4x and 5x. The smaller leg is 4/5 of the larger leg. Since the area is 320, ½(4x)(5x)=320, $10x^2 = 320$. Then $x^2 = 32$ and $x=\sqrt{32}=4\sqrt{2}$. The legs will have lengths $16\sqrt{2}$ and $20\sqrt{2}$.

17. $w^2 = 18^2 + 30^2 + 20^2 = 324+900+400=1624$. Thus, $w = \sqrt{1624} = 2\sqrt{406}$ m. This is slightly more than 40 m. (See figure 3.38)

18.   (See figure 3.39) Let the larger square have sides of length 2x. Thus its area is $4x^2$. The side of the smaller square is the hypotenuse of a right triangle with both legs of length x. Thus, $s^2 = 2x^2$ and $s = x\sqrt{2}$. The area of the smaller square is A = $(x\sqrt{2})^2 = 2x^2$. This is ½ of the area of the larger square.

19.   In this diagram, it is clear that the fielder's throw is along the diagonal of a rectangle with sides 120 ft. and 90 ft.

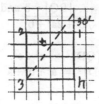

        Since 90 and 120 are multiples of 3 and 4, we obtain t = 30(5) = 150 so the throw would be 150 ft.

20.   Let the side of the square have length x. Then, $x^2 = 169$ so $x=13$. The diagonal has length d such that $d^2 = 2(13)^2$ so $d = 13\sqrt{2}$.

21. $d = 8\sqrt{2}$ so $128 = s^2 + s^2 = 2s^2$. Thus, $s^2=64$ = area of the square.

22. $d = 8$ so $64 = 2s^2$ and $s^2 = 32$ = Area of the square.

23.   (a)   Let the sides of the screen have lengths 3x and 4x. Then,
            $d^2 = 25^2 = 625 = (3x)^2 + (4x)^2 = 25x^2$.
            $x^2 = 25$ and $x=5$. The sides are 15 and 20 so the viewing area of the screen is $(15)(20)=300$ sq. cm.

       (b) $d^2 = 50^2 = 2500 = 25x^2$. $x^2 = 100$ and $x=10$. The sides are 30 and 40 so the viewing area is $(30)(40)=1200$ which is four times the viewing area in part (a).

24.   In figure 3.40, construct a line through S and parallel to side PQ intersecting side $\overline{RQ}$ at point T. $|ST| = |PQ|$.

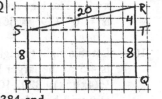

$20^2 = 4^2 + |ST|^2$ so $|ST|^2 = 400 - 16 = 384$ and $|ST| = \sqrt{384} = 8\sqrt{6}$. Thus, $|PQ| = 8\sqrt{6}$ units.

25.   Given two right triangles, $\triangle ABC$ and $\triangle XYZ$, such that $|AB| = |XY|$ and the hypotenuse $\overline{AC}$ of $\triangle ABC$ and the hypotenuse $\overline{XZ}$ of $\triangle XYZ$ have the same length. Prove: $|BC| = |YZ|$

$|AB|^2 + |BC|^2 = |AC|^2$ and $|XY|^2 + |YZ|^2 = |XZ|^2$. Since $|AB| = |XY|$ and $|AC| = |XZ|$ by substitution we obtain $|AB|^2 + |YZ|^2 = |AC|^2$. Thus, $|AB|^2 + |BC|^2 = |AB|^2 + |YZ|^2$. By subtraction $|BC|^2 = |YZ|^2$ Therefore, $|BC| = |YZ|$

26.   (a)   Given rectangle ABCD with right triangle AQD inscribed in it such that $|AB|=a$, $|BQ|=b$ and $|QC|=c$. (See fig. 3.41) Prove $|AD| = \sqrt{b^2+2a^2+b^2}$

$|AQ|^2 = a^2+b^2$ and $|QD|^2 = a^2+c^2$. Therefore,
$|AD|^2 = |AQ|^2+|QD|^2 = a^2+b^2+a^2+c^2 = b^2+2a^2+c^2$. Thus
$|AD| = \sqrt{b^2+2a^2+c^2}$.

       (b) $|AD|^2 =(b+c)^2 = b^2+2bc+c^2 = b^2+2a^2+c^2$, by part (a). By subtraction, $2bc=2a^2$ so $a^2=bc$.

27. Given $\triangle ABC$ with $|AC|=|CB|$ and $\overline{CN} \perp \overline{AB}$.
Prove $|AN|=|NB|$. (See fig. 3.42)

$|AC|^2 = |CN|^2+|AN|^2$ and $|CB|^2 = |CN|^2+|NB|^2$. Since
$|AC| = |CB|$, by substitution we have
$|CN|^2+|AN|^2=|CN|^2+|NB|^2$ thus $|AN|^2=|NB|^2$ and $|AN|=|NB|$.

28. Given $\triangle ABC$ is a right triangle with the right angle at C and point P on
$\overline{AB}$ the foot of a perpendicular from C. (see fig. 3.43)

(a) Prove: $|PC|^2 = |AP||PB|$

Let $|AC|=b$, $|CB|=a$, $|AP|=x$, $|PB|=y$ and $|PC|=h$. By
Pythagoras, $a^2+b^2=(x+y)^2=x^2+2xy+y^2$ while $a^2=h^2+y^2$ and $b^2=h^2+x^2$.
By substitution, $2h^2+x^2+y^2=x^2+2xy+y^2$. Therefore $2h^2=2xy$ and
$|PC|^2=|AP||PB|$.

(b) Prove: $|AC|^2=|AP||AB|$

Using the notation above to prove this statement we need to show
$b^2=x(x+y)$. From the above statements we use
$a^2+b^2=x^2+2xy+y^2$ and
$a^2=h^2+y^2$. Subtracting the second equation from the first we obtain
$b^2=x^2+2xy-h^2$. By adding $b^2=h^2+x^2$ to each side
$2b^2=2x^2+2xy$. By dividing by 2 and factoring the right side we note
$b^2=x(x+y)$ and $|AC|^2=|AP||AB|$.

29. Given figure 3.44 with $|PM|=|MQ|$ and $|XM'|=|M'Y|$ and line L
perpendicular to $\overline{PQ}$ and $\overline{XY}$.

(a) Prove: $|PX|=|QY|$

Draw a perpendicular from P to P' on $\overline{XY}$ and another from Q to Q' on
$\overline{XY}$. (See fig. 3.45). Since PQQ'P' is a rectangle $|PQ|=|P'Q'|$ and with
M the midpoint of $\overline{PQ}$ we also have M' the midpoint of $\overline{P'Q'}$. Thus
$|P'M'|=|M'Q'|$ and by subtraction we obtain
$|XP'|=|YQ'|$. In the right triangles $\triangle XPP'$ and $\triangle YQQ'$,
$|PP'|=|QQ'|$ and $|XP'|=|YQ'|$. By the Pythogoras Theorem,
$|PP'|^2 + |XP'|^2 = |PX|^2$ and
$|QQ'|^2 + |YQ'|^2 = |QY|^2$. Since the left sides of the equations are
equal we have $|PX|^2=|QY|^2$ so $|PX|=|QY|$.

(b)   Prove: $|PY| = |QX|$

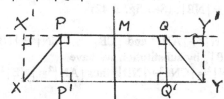

Following the approach of part (a), construct a perpendicular from X meeting line $\overline{PQ}$ at point X'. Construct a perpendicular from Y meeting line $\overline{PQ}$ at Y'. XX'PP' is a rectangle thus $|XP'| = |X'P|$ and $|PP'| = |XX'|$. YY'QQ' is also a rectangle thus $|QY'| = Q'Y|$ and $|QQ'| = |YY'|$. From part (a) we have $|XP'| = |Q'Y|$ and $|PQ| = |P'Q'|$. Now one can show that $|XP'| = |P'Y|$.

$$\begin{aligned}
|X'P| &= |PQ| + |X'P| \quad \text{SEQ} \\
&= |P'Q'| + |XP'| \quad \text{by substitution} \\
&= |P'Q'| + |Q'Y| \quad \text{by substitution} \\
&= |P'Y|.
\end{aligned}$$

Since, $|XX'| = |YY'|$ and $|X'P| = |P'Y|$ the right triangles PY'Y and QX'X are congruent by SAS. Therefore, by corresponding parts of these congruent triangles, $|PY| = |QX|$.

30.   Let the right triangle ABC have c as the length of the hypotenuse and a and b as the lengths of its legs.
       Prove: $c > a$

       By Pythagoras $c^2 = a^2 + b^2$. We can note that $c^2 > a^2$ since $b^2 > 0$. Thus, $c > a$ since length is non-negative. Therefore, the length of the hypotenuse of a right triangle is greater than or equal to the length of a leg.

31.   (a)   Given $|AF| = 10$ in figure 3.46, the area of square ADEF is 100. $\triangle ADG$ has one fourth of this area, thus it is 25 sq. units.

       (b)   Given $|CE| = 18$. $|CD| = |BH| = |HG| = \frac{1}{2}|DE|$ so $|HG| = 6$. $|CD| = |HD| = 6$ so $|AD| = 12$.
              Area of $\triangle ADG = \frac{1}{2}(6)(12) = 36$ sq. units.

       (c)   Given $|BD| = 3\sqrt{2}$. The area of the square ABDG $= (3\sqrt{2})^2 = 18$. Area of $\triangle ADG = \frac{1}{2}(18) = 9$ sq. units.

       (d)   Given area of square BCDH $= 49$. $|BC|^2 = 49$ so $|BC| = 7 = |HD| = |AH| = |HG|$. Thus the area of $\triangle ADG = \frac{1}{2}(14)(7) = 49$ sq. units

(e)   Given area of AGDEF = 27. The area of AGDEF is one-fourth the area of square ADEF. Let A = area of square ADEF. Thus 27=(3/4)A so A=36. ▵ADG is one-fourth of the square ADEF so its area is (1/4)(36)=9 sq. units.

Experiment 3-3

1. $d(p,Q)=9$

2. $d(P,R)=9$

3.   ▵PQR is an isosceles right triangle.

4. $d(Q,R) = \sqrt{9^2+9^2} = \sqrt{162} = 9\sqrt{2}$.

5.   Let $Z=(-1,-2)$, then $d(X,Z)=9$, $d(Y,Z)=4$ so $d(X,Y)=\sqrt{9^2+4^2}=\sqrt{97}$.

6.   Let $Y=(x_1,3)$, then $d(A,Y)=\sqrt{(x_1-5)^2}$ and $d(X,Y)=\sqrt{(x_2-3)^2}$.
     $d(A,X)=\sqrt{(x_1-5)^2+(x_2-3)^2}$.

CHAPTER 4

The Distance Formula

§ 1. Exercises
(Pages 113,114)

1. $d^2 = (2-1)^2 + (1-5)^2 = 17$, $d = \sqrt{17}$

2. $\sqrt{26}$

3. $\sqrt{61}$

4. $\sqrt{2}$

5. $\sqrt{106}$

6. $\sqrt{106}$

7. $2\sqrt{2}$

8. $\sqrt{58}$

9. $2\sqrt{53}$

10. $\sqrt{5}$

11. $|AB| = 7$, $|DC| = 7$, $|AD| = \sqrt{29}$, $|BC| = \sqrt{29}$ ABCD is a parallelogram.

12. (a) B and C determine a horizontal line segment of length 8 units.

    (b) A and B determine a vertical line segment of length 6.

    (c) $|AC| = 10$

13. Let x= distance between (3,5) and (-1,2).
$x^2 = (3-(-1))^2 + (5-2)^2 = 25$, x=5.

Let y= distance between (3,5) and (3,0).
$y^2 = (3-3)^2 + (5-0)^2 = 25$, y=5. Thus x=y.

14. $25 = (2-(-1))^2 + (k-1)^2 = 9 + k^2 - 2k + 1 = k^2 - 2k + 10$
$0 = k^2 - 2k - 15 = (k-5)(k+3)$
Thus k=5 or k=-3.

15.

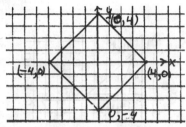

$s^2 = (0-4)^2 + (4-0)^2 = 16 + 16 = 32$, $s = 4\sqrt{2}$ units.

16. (a) $|AK| = \sqrt{(0-4)^2 + (y-3)^2} = \sqrt{16 + y^2 - 6y + 9}$
$|AK| = \sqrt{y^2 - 6y + 25}$

(b) $|BK| = \sqrt{(0+8)^2 + (y-0)^2} = \sqrt{y^2 + 64}$

(c) $|BK| = 2|AK|$ or $|BK|^2 = 4|AK|^2$
$y^2 + 64 = 4(y^2 - 6y + 25)$
$y^2 + 64 = 4y^2 - 24y + 100$
$0 = 3y^2 - 24y + 36$
$0 = y^2 - 8y + 12$
$0 = (y-6)(y-2)$, thus y=6 or y=2.

17. Let a = distance between (0,y) and (-3,2) and b = distance between (0,y) and (5,1).
$a^2 = (0+3)^2 + (y-2)^2 = 9 + y^2 - 4y + 4 = y^2 - 4y + 13$
$b^2 = (0-5)^2 + (y-1)^2 = 25 + y^2 - 2y + 1 = y^2 - 2y + 26$. Since $a^2 = b^2$,
$y^2 - 4y + 13 = y^2 - 2y + 26$
$-2y = 13$
$y = -13/2 = -6\frac{1}{2}$

18. Let P=(a,b) and Q=(x,y) and d(P,Q)=0.
Prove: P=Q

$d(P,Q) = \sqrt{(a-x)^2 + (b-y)^2} = 0$
$(a-x)^2 + (b-y)^2 = 0$. Since the sum of these squares equals 0, each term equals 0. Therefore, a-x=0 and a=x. Also b-y=0 so b=y. Thus P=Q.

19.　$[d(A,B)]^2 = (a_1-b_1)^2+(a_2-b_2)^2$
　　　$[d(rA,rB)]^2 = (ra_1-rb_1)^2+(ra_2-rb_2)^2$
　　　　　　　　$= r^2[(a_1-b_1)+(a_2-b_2)^2]$
　　　$d(rA,rB) = r\sqrt{(a_1-b_1)^2+(a_2-b_2)^2}$
　　　　　　　　$= rd(A,B)$

20.　Let $P=(k,6)$, $A=(1,1)$ and $B=(3,5)$.
　　　$(k-1)^2+(6-1)^2=(k-3)^2+(6-5)^2$
　　　$k^2-2k+1+25=k^2-6k+9+1$
　　　$4k=10-26=-16$
　　　$k=-4$

21.

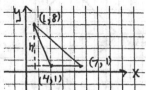

　　　$h=7$, $b=3$, $A=\frac{1}{2}(7)(3)=\frac{1}{2}(21)=10\frac{1}{2}$ sq. units

22.　$s=4$ so the area $= 16$ (Choice e)

§ 2. Exercises
(Page 119)

1.　(a)　$d(P,Q)^2 = (1-(-1))^2+(2-3)^2+(4-(-2))^2$
　　　　　　　　$= 2^2+(-1)^2+6^2$
　　　　　　　　$= 4+1+36$
　　　　　　　　$= 41$
　　　　　　$d(P,Q) = \sqrt{41}$
　　　(b)　$\sqrt{13}$
　　　(c)　$\sqrt{50} = 5\sqrt{2}$
　　　(d)　$\sqrt{70}$

2.　Let $A=(a_1,a_2,a_3)$ and $B=(b_1,b_2,b_3)$ and r, a positive real number.
　　Prove: $d(rA,rB)=rd(A,B)$

　　　$d(rA,rB)^2 = (ra_1-rb_1)^2+(ra_2-rb_2)^2+(ra_3-rb_3)^2$
　　　　　　　　$= r^2(a_1-b_1)^2+r^2(a_2-b_2)^2+r^2(a_3-b_3)^2$
　　　　　　　　$= r^2[(a_1-b_1)^2+(a_2-b_2)^2+(a_3-b_3)^2]$
　　　　　　　　$= r^2d(A,B)^2$
　　Therefore, $d(rA,rB)=rd(A,B)$

§ 3. Exercises
(Pages 121-122)

1. $(x+3)^2+(y-1)^2=4$

2. $(x-1)^2+(y-5)^2=9$

3. $(x+1)^2+(y+2)^2=1/9$

4. $(x+1)^2+(y-4)^2=4/25$

5. $(x-3)^2+(y-3)^2=2$

6. $x^2+y^2=8$

7.   Center $(1,2)$, radius$=5$

8.   Center $(-7,3)$, radius$=\sqrt{2}$

9.   Center $(-1,9)$, radius$=\sqrt{8}=2\sqrt{2}$

10.   Center $(-1,0)$, radius$=\sqrt{(5/3)}=(\sqrt{15})/3$

11.   Center $(5,0)$, radius$=\sqrt{10}$

12.   Center$(0,2)$, radius$=\sqrt{(3/2)}=(\sqrt{6})/2$

13.   Let $O=(3,5)$ be the center and $A=(0,1)$ be a point on the circle.

(a) $r^2 = (3-0)^2+(5-1)^2 = 9+16 = 25$, radius$=5$

(b) $(x-3)^2+(y-5)^2=25$. A point on the y-axis has the x-coordinate 0. Thus $9+(y-5)^2=25$ and $(y-5)^2=16$ so $(y-5)=4$ or -4. $y=1$ or 9. The other point of intersection of the circle with the y-axis is $(0,9)$.

(c)   On the x-axis all points have their y-coordinate equal to 0. Thus $(x-3)^2+25=25$ and $(x-3)^2=0$. Hence, x-3=0 and x=3. The only intersection of the circle with the x=axis is $(3,0)$.

14. $(x-3)^2+(y-5)^2=25$. Let $P=(4,y)$.
$(4-3)^2+(y-5)^2=25$
$1+(y-5)^2=25$
$(y-5)^2=24$
$y-5=\sqrt{24}$ or $5+\sqrt{24}$
$y=5+\sqrt{24}$ or $5-\sqrt{24}$
Therefore the points are $(4,5+\sqrt{24})$ and $(4,5-\sqrt{24})$.

15.   (a)   All points within 2 units of the origin.
Points in the interior of the circle with center $(0,0)$ and radius 2.

(b)   Only one point, $(8,1)$

16.   (a) $x^2+y^2+z^2=1$

(b) $x^2+y^2+z^2=9$

(c) $x^2+y^2+z^2=r^2$

(d)   (i) $(x-1)^2+(y+3)^2+(z-2)^2=1$

(ii) $(x-1)^2+(y-2)^2+(z+5)^2=1$

(iii) $(x+1)^2+(y-5)^2+(z-3)^2=4$

(iv) $(x+2)^2+(y+1)^2+(z+3)^2=4$

(v) $(x+1)^2+(y-1)^2+(z-4)^2=9$

(vi) $(x-1)^2+(y-3)^2+(z-1)^2=49$

17.   (a)   Center $(0,1,-2)$, radius $=4$

(b)   Center $(-3,2,0)$, radius $=\sqrt{7}$

(c)   Center $(-5,0,0)$, radius $=\sqrt{2}$

(d)   Center $(2,0,-1)$, radius $=\sqrt{8}=2\sqrt{2}$

CHAPTER 5

Some Applications of Right Triangles

§ 1. Exercises
(Pages 129-132)

1. See fig. 5.7
    a=10 cm, b=4 cm, c=7 cm

2. Line K is the perpendicular bisector of $\overline{PQ}$ and line L is the perpendicular bisector of QM. (see fig. 5.8)
    Prove: d(O,P)=d(O,M)

    d(O,P) = d(O,Q) by hypothesis and Thm. 5.1 applied to K = d(O,M) by hypothesis and Thm. 5.1 applied to L

3. Since |AX| = |AY|, A is on the perpendicular bisector of $\overline{XY}$. There is only one line through A and perpendicular to $\overline{XY}$ so L must be the perpendicular bisector of $\overline{XY}$.

4. No, since $\overline{XY}$ would not necessarily intersect $\overline{AB}$ at its midpoint.

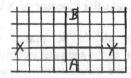

5. Let MNPQ be a rhombus. (see fig. 5.10)
    Prove $\overline{PM}$ and $\overline{QN}$ are perpendicular bisectors of each other.

    |NP| = |NM| = |PQ| = |QM| since they are sides of a rhombus. Therefore, since N and Q are equidistant from P and M and two points

determine a line, $\overline{QN}$ is the perpendicular bisector of $\overline{PM}$. Similarly, M and P are equidistant from N and Q so $\overline{PM}$ is the perpendicular bisector of $\overline{QN}$.

6.   In figure 5.10, $|PM|=24$ and $|NQ|=18$. Let O be the intersection point for the diagonals. Since they bisect each other, $|OP|=12$ and $|OQ|=9$. In the right triangle OPQ, $|PQ|^2=9^2+12^2=81+144=225$. Therefore $|PQ|=15$. All sides equal 15 units.

7.   In figure 5.10, let O be the point of intersection of the diagonals. $|PM|=16$ and $|NQ|=12$. Thus $|OP|=8$ and $|OQ|=6$ and $|PQ|^2=6^2+8^2=36+64=100$. Therefore, each side equals 10 units.

8.   Given $\overline{AC}$ is on the perpendicular bisector of $\overline{BD}$ and $\overline{BD}$ is on the perpendicular bisector of $\overline{AC}$.
      Prove: ABCD is a rhombus.

      Since $\overline{AC}$ is on the perpendicular bisector of $\overline{BD}$,
      $d(A,D)=d(A,B)$ and
      $d(C,D)=d(B,C)$. We also note that
      $d(A,B)=d(B,C)$ and
      $d(A,D)=d(C,D)$ since $\overline{BD}$ is on the perpendicular bisector of $\overline{AC}$. From these equations we conclude that $d(A,D)=d(C,D)=d(C,B)=d(A,B)$. Therefore, ABCD is a parallelogram and a rhombus.

9.   In figure 5.11, $d(X,A)=d(Y,A)$ and $d(X,B)=d(Y,B)$.
      Prove: Line L_{AB} is the perpendicular bisector of XY.

      By the perpendicular bisector theorem and the given, A lies on the perpendicular bisector of $\overline{XY}$. Also for the same reasons B lies on the perpendicular bisector of $\overline{XY}$. Since points A and B determine a unique line, L_{AB} is the perpendicular bisector of $\overline{XY}$.

10.   The receiving tower should be placed at the intersection of the perpendicular bisector of the segment joining YGLEPH and ZYZZX and the circle centered at the transmitter with radius 100 KM.

11.   Form the $\triangle PQR$ and construct the perpendicular bisector of each side. The intersection of these 3 lines is the point equidistant from P, Q, and R. This is the only point meeting these conditions. This point is called the circumcenter of the triangle and is the center of the circle passing through the 3 vertices.

12.   See figure 5.3(b). SEG indicates that $d(P,Q)+d(Q,X)=d(P,X)$. Since $d(P,X)=d(Q,X)$ we can cancel and obtain $d(P,Q)=0$. This again

contradicts the fact that P and Q are distinct points so this arrangement is not possible. Thus X lies on the segment PQ.

13. Build the dock at the point where the shoreline intersects the perpendicular bisector of the segment joining Factory A with Factory B.

14. In figure 5.13, P and Q are points on the circle centered at O.
     Prove: The perpendicular bisector of the chord, $\overline{PQ}$, passes through O.

     Since P and Q lie on the circle, they are equidistant from O. The perpendicular bisector of $\overline{PQ}$ includes all points equidistant from P and Q so must include O. Therefore, the perpendicular bisector of $\overline{PQ}$ passes through O.

15. In figure 5.13, let the line L through O be perpendicular to $\overline{PQ}$ at M.
     Prove: Line L bisects $\overline{PQ}$.

     Since P and Q are on the circle, $d(O,P)=d(O,Q)$. Therefore, O is on the perpendicular bisector of $\overline{PQ}$. Since there is only one line passing through O and perpendicular to $\overline{PQ}$, L must be the perpendicular bisector of $\overline{PQ}$. Thus line L bisects the chord, $\overline{PQ}$.

16. In figure 5.14, let X and Y be the points of intersection of the two circles centered at P and Q.
     Prove: $\overline{PQ}$ passes through the midpoint of $\overleftrightarrow{XY}$.

     Since X and Y are on the circle centered at P, $d(P,X)=d(P,Y)$. Therefore P lies on the perpendicular bisector of $\overline{XY}$. Since X and Y are on the circle centered at Q, Q lies on the perpendicular bisector of $\overline{XY}$. These two points, P and Q, determine a unique line, the perpendicular bisector of $\overline{XY}$, thus $\overline{PQ}$ contains the midpoint of $\overline{XY}$.

§ 2. Exercises
(Pages 143-147)

1.   (a) $m(\angle A)=40$, $m(\angle B)=m(\angle C)=\frac{1}{2}(180-40)=70°$

     (b) $m(\angle B)=m(\angle C)=60°$

     (c) $m(\angle B)=m(\angle C)=45°$

     (d) $m(\angle B)=m(\angle C)=35°$

2.   (a) $a=5$, $b=3$, $h=$ height

$5^2=h^2+(3/2)^2$

$25=h^2+9/4$

$h^2=25-9/4=100/4-9/4=91/4$

$h=(\sqrt{91})/2$

Area of $\triangle ABC=\frac{1}{2}(\sqrt{(91/2)})(3)=(3/4)\sqrt{91}$ sq. units

       (b) $a=10$, $b=8$

$10^2=h^2+4^2$

$100=h^2+16$

$h^2=100-16=84$

$h=\sqrt{84}=2\sqrt{21}$

Area of $\triangle ABC=\frac{1}{2}(8)(2\sqrt{21})=8\sqrt{21}$ sq. units

       (c) $a=5$, $b=2$

$5^2=h^2+1^2$

$25=h^2+1$

$h^2=25-1=24$

$h=\sqrt{24}=2\sqrt{6}$

Area of $\triangle ABC=\frac{1}{2}(2)(2\sqrt{6})=2\sqrt{6}$ sq. units

3.   Area of sides $= 2(10)(3)+2(4)(3)=84$

Area of flat roof $= 2(10)(3)=60$

$\triangle$ areas $= 2[\frac{1}{2}(4)(\sqrt{5})]=4\sqrt{5}$

Total Area$=144+4\sqrt{5}$ sq. units

4.   (a)   Let $h=$ altitude of the equilateral triangle. By the Pythagoras Theorem, $h^2+(6/2)^2=6^2$ so $h=3\sqrt{3}$. In general, $h=(s/2)\sqrt{3}$. Area$=\frac{1}{2}(6)(3\sqrt{3})=9\sqrt{3}$ sq. m

       (b)   Area$=\frac{1}{2}(3)(3\sqrt{3}/2)=9\sqrt{3}/4$ sq m

       (c)   Area$=\frac{1}{2}(\sqrt{3})(3/2)=3\sqrt{3}/4$ sq. m

5.

$h^2=s^2-(s/2)^2$

$h^2=3s^2/4$

$h=s(\sqrt{3})/2$

6.   Area$=\frac{1}{2}(s)(s\sqrt{3})/2=(s^2\sqrt{3})/4$

7.   Perimeter$=12$ m, $s=4$ m.

$A=16\sqrt{3}/4=4\sqrt{3}$ sq. m

8.   Perimeter=24 m, s=8 m.
      $A=64\sqrt{3}/4=16\sqrt{3}$ sq. m.

9.   Area = $12+16\sqrt{3}/4=12+4\sqrt{3}$ sq. m. (see fig.5.39)
      Coverage per can = 9 sq m.
      Number of cans=$(12+4\sqrt{3})/9=2.103$ cans. Thus 3 cans are needed.
      (3 cans)($4.50 per can)=$13.50 cost for this project.

10.   (a) $x=180-(2)(65)=50°$ (see fig. 5.40)

      (b) $x=180-(2)(50)=80°$

11.   (a)   In figure 5.41, $|MP|^2=1^2+1^2=2$ so $|MP|=\sqrt{2}=|PQ|$.
            $|MQ|^2=1^2+(1+\sqrt{2})^2=1+1+2\sqrt{2}+2=4+2\sqrt{2}$

      (b)   m(angle MPQ)=$180-45=135$
            m(angle Q)=$\frac{1}{2}(180-135)=22\frac{1}{2}°$

      (c)   m(angle QMO)=$45+22\frac{1}{2}=67\frac{1}{2}°$

12. $s=8$
      Area = $64\sqrt{3}/4=16\sqrt{3}$ sq. units

13. $x^2=s^2+s^2=2s^2$
      $s^2=x^2/2$
      $s=x/\sqrt{2}$

14.   In figure 5.43, let $\triangle$PMQ be isosceles with $|PM|=|MQ|$. Let X and Y
      lie on $\overline{PQ}$ so that $d(P,X)=d(Y,Q)$.
      Prove: $\triangle$XMY is isosceles.

      In $\triangle$PMQ draw the altitude from vertex M intersecting $\overline{PQ}$ at point D.
      In an isosceles triangle this altitude bisects the base so $d(P,D)=d(D,Q)$.
      Let $d(P,X)=d(Y,Q)=a$. By the SEQ postulate
      $d(P,D)=a+d(X,D)$ and
      $d(D,Q)=a+d(D,Y)$. By the above equation
      $a+d(X,D)=a+d(D,Y)$. By subtraction,
      $d(X,D)=d(D,Y)$. Therefore D is the midpoint of $\overline{XY}$. Thus the altitude
      through M is also the perpendicular bisector of $\overline{XY}$. As a result,
      $d(X,M)=d(Y,M)$ and $\triangle$XMY is isosceles.

15.   Let $\triangle$XYZ have sides of length s and $\triangle$ABC have sides of length 2s.
      Then Area of $\triangle$XYZ = $s^2\sqrt{3}/4$. The area of $\triangle$ABC = $(2s)^2\sqrt{3}/4=s\sqrt{3}$.
      This is 4 times the area of $\triangle$XYZ.

16.   Given ΔABC with |AB|=|AC|.
       Prove m(∠x)=m(∠y). (See fig. 5.44)

       By the Isosceles Triangle Theorem m(∠ABC)=m(∠BCA). ∠ABC and
       ∠x form a linear pair so m(∠ABC)+m(∠x)=180. Also ∠BCA and
       ∠y form a linear pair and m(∠BCA)+m(∠y)=180. Since both sums
       equal 180 they equal each other and by substituting m(∠ABC) for
       m(∠BCA) we obtain m(∠ABC)+m(∠x)=m(∠ABC)+m(∠y). By
       subtraction we have m(∠x)=m(∠y).

17.   In figure 5.45, |BC|=|CA|=|AD| and m(∠B)=x°.
       Prove: m(∠EAD)=3m(∠B)

       For ΔABC, which is isosceles, we have m(∠A)=m(∠B) by the Isosceles
       Triangle Theorem. ∠DCA is an exterior angle of ΔABC. Thus, by the
       Exterior   Angle   Theorem   (p.58,   17)   we   conclude   that
       m(∠DCA)=m(∠A)+m(∠B)=2x°. Because |CA|=|AD|, ΔACD is
       isosceles   and,   by   the   Isosceles   Triangle   Theorem,   we   get
       m(∠D)=m(∠DCA)=2x°. Since ∠EAD is an exterior angle for ΔABD
       we   have   m(∠EAD)=m(∠B)+m(∠D)=x+2x=3x°.   Therefore,
       m(∠EAD)=3m(∠B).

18.   Given ΔQED has |QE|=|ED| and segment IT parallel to the base QD.
       Prove: m(∠x)=m(∠y) (See fig. 5.46)

       Since |QE|=|ED| in ΔQED, m(∠Q)=m(∠D) by the Isosceles Triangle
       Theorem. Since IT is parallel to QD the parallel angles have equal
       measure so m(∠Q)=m(∠x) and m(∠D)=m(∠y). By substitution we
       have m(∠x)=m(∠y).

19.   Let ΔABC be a right triangle with legs of lengths a and b and hypotenuse
       of length c. Prove: The length c is greater than a or b.

       By the Pythagoras equation $c^2=a^2+b^2$. Suppose b≥c. Then $0≥a^2$
       contradicting the fact that a is the length of a side of this triangle.
       Therefore, c cannot equal b. Thus the hypotenuse is greater in length than
       either leg.

20.   Given ΔABC where |AB|=|BC| and line L passing through B and
       perpendicular to AC at point O. (See fig. 5.47)
       Prove: Line L bisects ∠ABC.

       Since |AB|=|BC| vertex B is on the perpendicular bisector of AC by
       the Perpendicular Bisector Theorem. Since L contains B and is
       perpendicular to AC we know that L is the perpendicular bisector of AC.

Thus O is the midpoint of AC and $|AO|=|OC|$. For the right triangles $\triangle ABO$ and $\triangle BOC$ we have $|AO|=|OC|$ and $|OB|=OB|$. By the RT Theorem (p. 47) we get m($\angle ABO$)=m($\angle OBC$) and L bisects $\angle ABC$.

21.   Given the construction of the angle bisector on p. 25, prove the construction produces the angle bisector.

By the first step of the construction, P and Q are marked on the legs of the $\angle BAC$ at an equal distance from A. Draw the segment $\overline{PQ}$. The second part of the construction obtains a line perpendicular to $\overline{PQ}$ through A. By the result in problem 20 we know that this line bisects $\angle BAC$.

Construction 5-3
(Page 148)

(a)   Bisect a right angle.

(b)   Construct an equilateral triangle.

(c)   Bisect the 60° angle.

(d)   Bisect the 30° angle.

For 75° combine 60° and 15°.

For 105° combine 90° and 15° or 60° and 45°.

For 135° combine 90° and 45°.

For 120° combine 60° and 60°.

§ 3. Exercises
(Pages 156-161)

1.   (a) $x=\frac{1}{2}(75)=37\frac{1}{2}$

      (b) $x=\frac{1}{2}(360-110)=\frac{1}{2}(250)=125$

      (c) $x=\frac{1}{2}(180)=90$

      (d) $x=\frac{1}{2}(180-90)=45$

      (e) $\triangle COB$ is equilateral so m($\angle COB$)=60°. $x=\frac{1}{2}(60)=30$.

2.   The sum of the measures of these arcs is 360°.

3.   In figure 5.63 we are given that $\overline{PQ}$ is a diameter and M is any other point on the circle.
Prove: In ΔPMQ, ∠M is a right angle.

From Theorem 5-4 and that ∠M of ΔMPQ is inscribed in the semi-circle determined by the diameter PQ, we have m(∠M)=90° so ∠M is a right angle.

4.   In figure 5.64, draw the radius perpendicular to $\overline{AB}$ meeting $\overline{AB}$ at M. For the right triangle AMO we have |OA|=10 cm and |OM|=8 cm. |AM|=½|AB| and $10^2=8^2+|AM|^2$. So 100-64=36=|AM|². Thus |AM|=6 and |AB|=12 cm.

5.   By the same reasoning we have 100-9=91=|AM|². So |AM|=√91 and |AB|=2√91 cm.

6.   In figure 5.65, d(O,P)=15, line L is tangent to the circle and the length of the radius of the circle is 12. By the definition of a tangent ∠OQP is a right angle. Thus we have in ΔOPQ $15^2=12^2+|PQ|^2$. So |PQ|²=225-144=81 and |PQ|=9.

7.   In figure 5.66 in the smaller circle draw the radius to the point of tangency, T. Then |OT|=3, |OB|=7 and |BT|²+3²=7². So |BT|²=49-9=40 and |BT|=√40=2√10. |AB|=2|BT|=4√10.

8.   In figure 5.67, line L is tangent to the circle at P.
Prove: No other point on L can intersect the circle.

Let Q be a point on L different from P.. Since Q is on L and P is the point where the perpendicular from O meets L we know by Theorem 3.5 (page 100) that |OQ| > |OP| so Q lies outside of the circle. Therefore P, the point of tangency, is the only point of L in common with the circle.

9.   Let two lines through point M be tangent to a circle at points P and Q. Prove: |PM| = |QM|.

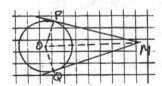

For radii of this circle, let $|OP| = |OQ| = r$ and let $|OM| = y$. Since P and Q are points of tangency for the two lines, the triangles $\triangle OPM$ and $\triangle OQM$ are right triangles. By Pythagoras we have
$|PM|^2 = y^2 - r^2 = |QM|^2$.
Thus $|PM| = |QM|$.

10. In figure 5.69 we are given a quadrilateral whose sides are tangent to the circle.
Prove: $|OV| + |LE| = |LO| + |VE|$.

By the proof in problem 9 we have
$|OT| = |OA|$
$|LM| = |LA|$
$|EM| = |EH|$
$|VT| = |VH|$. If we add these equations we obtain
$|OT| + |LM| + |EM| + |VT| = |OA| + |LA| + |EH| + |VH|$.
By a rearrangement of these terms we have
$(|OT| + |VT|) + (|LM| + |EM|) = (|OA| + |LA|) + (|EH| + |VH|)$.
By the SEG Postulate we can replace these pairs by
$|OV| + |LE| = |LO| + |EV|$.

11. In figure 5.70 we are given that all three lines are tangent to the circle and $|RP| = 5$ cm.
$|RP| = |TR| = 5$. $|PQ| = |QE|$ and $|TS| = |ES|$.
$5 = |RP| = |RQ| + |QP| = |RQ| + |QE|$ and
$5 = |TR| = |TS| + |SR| = |ES| + |SR|$.   Since $|QE| + |ES| = |QS|$,
$|RQ| + |QS| + |SR| = 10$. Therefore, the perimeter of $\angle QRS$ is 10 cm.

12. In figure 5.71 we are given a tangent to the circle from point M and a secant to the circle from M and passing through the center, O.
Prove: $|PM|^2 = |MQ||MR|$.

Draw $\overline{OP}$ and let $c = |OP| = |OR| = |OQ|$. Also let $a = |PM|$ and $b = |MR|$. Since radius $\overline{OP}$ is perpendicular to $\overline{PM}$, in the right triangle OPM we get
$(b+c)^2 = a^2 + c^2$ and $b^2 + 2bc + c^2 = a^2 + c^2$. By cancellation we obtain $a^2 = b^2 + 2bc = b(b+2c)$. By substitution we have
$|PM|^2 = |MR||MQ|$ since $b + 2c = |MR| + |OR| + |OQ| = |MQ|$.

13. Let PUNT be a quadrilateral inscribed in a circle. (See fig. 5.72)
Prove: The opposite angles of PUNT are supplementary.

The angles of the quadrilateral PUNT are inscribed angles of the circle. Consequently,

m($\angle$P)=½m(arc UNT) and
m($\angle$N)=½m(arc TPU). Furthermore,
m($\angle$P)+m($\angle$N)=½m(arc UNT)+½m(arc TPU)
$$=½[m(arc\ UNT)+m(arc\ TPU)]$$
$$=½(360)=180°.$$
Therefore, $\angle$P and $\angle$N are supplementary. Since a quadrilateral's angles have measures adding up to 360^2 the remaining two angles, $\angle$U and $\angle$T, are supplementary.

14. In figure 5.73 we are given the chords of a circle meet at point O, not necessarily the center.
Prove: m($\angle$ROP)=½(sum of the measures of the arcs $\overarc{RP}$ and $\overarc{QS}$).

Draw $\overline{PS}$. $\angle$ROP is an exterior angle of $\triangle$POS. By the Exterior Angle Theorem, m($\angle$ROP)=m($\angle$S)+M($\angle$P). Since $\angle$S and $\angle$P are inscribed angles of the circle,
m($\angle$S)=½m(arc $\overarc{RP}$) and
m($\angle$P)=½m(arc $\overarc{QS}$). By substitution we have
m($\angle$ROP)=½m(arc $\overarc{PR}$)+½m(arc $\overarc{QS}$)
$$=½[m(arc\ \overarc{PR})+m(arc\ \overarc{QS})]$$

15. In figure 5.74 we are given R_{AC} and R_{AE} intersecting the circle.
Prove: m($\angle$A)=½[m(arc $\overarc{CE}$)-m(arc $\overarc{BD}$)]

Draw $\overline{CD}$. $\angle$CDE is an exterior angle for $\triangle$ADC. By the Exterior Angle Theorem, m($\angle$CDE)=m($\angle$A)+m($\angle$C). Solving for m($\angle$A), m($\angle$A)=m($\angle$CDE)-m($\angle$C). Since $\angle$CDE and $\angle$C are inscribed angles we have
m($\angle$CDE)=½m(arc $\overarc{CE}$) and
m($\angle$C)=½m(arc $\overarc{BD}$). By substitution we obtain
m($\angle$A)=½m(arc $\overarc{CE}$)-½m(arc $\overarc{BD}$) or
m($\angle$A)=½[m(arc $\overarc{CE}$)-m(arc $\overarc{BD}$)].

CHAPTER 6

Polygons

§ 1. Exercises

(Pages 174-176)

1.   (a)   yes, concave

      (b)   no

      (c)   yes, convex

      (d)   no

      (e)   yes, concave

2.   (a)   For the octagon $(8-2)180 = 1080°$

      (b)   For the pentagon $(5-2)180 = 540°$

      (c)   For the 12-gon $(12-2)180 = 1800°$

3.   (a) $(6-2)180/6 = 120°$

      (b) $(11-2)180/11 = 147 \ 3/11°$

      (c) $(14-2)180/14 = 154 \ 2/7°$

      (d) $(n-2)180/n$

4.   Let E be the measure of each exterior angle of this regular polygon. If each interior angle contains 153°, $E = 180-153 = 27°$. Since the sum of the exterior angles is 360° each one contains 360/n°.
Thus $27 = 360/n$ and $n = 360/27 = 13 \ 1/3$. No polygon can have a fractional number of sides. Thus an interior angle of 153° is impossible.

53

5.   (n-2)180/n=165 or use E=180-165=15. n=360/15=24 sides.

6.   (a)   (n-2)180=2700, n-2=15, n=17

      (b)   (n-2)180=1080, n-2=6, n=8

      (c)   (n-2)180=d, n-2=d/180, n=2+d/180

7.   If I=140°, E=180-140=40. n=360/40=9 sides

8.   (a)   rhombus

      (b)   rectangle

9.   The two equal sides could be of lengths 4 or 10. However, 4+4 is less than 10, the base has to be 4 and the equal sides 10 so the perimeter is 24.

10.   (a)   Since two of these equations have to be equal if the triangle is isosceles we let 2n-1=n+5 and find n=6 and the side lengths are 10, 11 and 11.

      (b)   Try the other two combinations.
             2n-1=3n-8, then n=7 and the sides have lengths 12, 13 and 13.
             3n-8=n+5, then n=6.5 and the side lengths are 11.5, 11.5 and

             Therefore, three values produce an isosceles triangle, 6, 7, 6.5.

      (c)   No value of n produces an equilateral triangle.

11.   P=36 so s=12. A=$12^2\sqrt{3}/4$=$36\sqrt{3}$ sq. units

12.   P=36 so s=9. A=9^2=81 sq. units

13.   Let this polygon have n sides. (See fig. 6.18)
      Prove: Sum of the outside angles equals (n+2)180°.

      One of these outside angles and the corresponding interior angle have a sum of measures equal to 360°. The total of the measures of these pairs of angles is n(360)°. The sum of the interior angles is (n-2)180 so the sum of the outside angles equals
      n(360)-(n-2)180
      =2n(180)-(n-2)180
      =(2n-(n-2))180
      =(n+2)180.

14. For figure 6.19, $P=2(5-2x)+2(4-2x)+8x=10-4x+8-4x+8x=18$ where the second term requires that $x<2$ (note that one side has length 0 if $x=2$.)

15. $A=(5)(4)-4x^2$

16. Let the regular hexagon have sides of length s.
    Prove: $A=3s^2\sqrt{3}/2$

    Draw the radii to the vertices of the hexagon. This produces six equilateral triangles with sides of length s. Thus the area of the hexagon can be computed by $A=6(s^2\sqrt{3}/4)=3s^2\sqrt{3}/2$.

17. $P=30$ cm. so $s=30/6=5$ cm. Thus the area $=3(5^2)\sqrt{3}/2=75\sqrt{3}/2$ sq. cm.

18. $A=3(2^2)\sqrt{3}/2=6\sqrt{3}$ sq. units

19. In figure 6.20 we have a regular hexagon with sides of length s and the perpendicular distance from the center to the side equal to x.
    Prove: $A=3xs$.

    Draw the radii to vertices F and E. The area of this equilateral triangle is ½xs. The total area of the hexagon equals $6(½xs)=3xs$.

CHAPTER 7

Congruent Triangles

§ 1. Exercises
(Pages 188-191)

1. $|AB| = |PQ|$, $|BC| = |QR|$, $|AC| = |PR|$, $m(\angle A) = m(\angle P)$, $m(\angle B) = m(\angle Q)$ and $m(\angle C) = m(\angle R)$.

2. (a) Yes, SSS

    (b) No, two possible triangles.

    (c) No, could be any size of the same shape.

    (d) Yes, ASA

    (e) No, The equal sides are not both between the given angles.

    (f) Yes, SAS

3. We are given that $\triangle ABC$ has $|AB| = |BC|$ and segment $\overline{BN}$ perpendicular to $\overline{AC}$. Prove: $\overline{BN}$ bisects $\angle ABC$.

    By the Isosceles Triangle Theorem, $m(\angle A) = m(\angle C)$. Also we have two right angles, $\angle ANB$ and $\angle CNB$. Since $|AB| = |BC|$ we have the ASA conditions met for congruence of $\triangle ANB$ and $\triangle CNB$. $m(\angle ABN) = m(CBN)$ since they are corresponding parts of these congruent triangles. Thus $\overline{BN}$ bisects $\angle ABC$.

4. We are given two right triangles, $\triangle ABC$ and $\triangle XYZ$, with right angles at B and Y and $|BC| = |YZ|$ and $m(\angle A) = m(\angle X)$.
    Prove: $\triangle ABC \simeq \triangle XYZ$

Since $\angle B$ and $\angle Y$ are right angle we have m($\angle B$)=m($\angle Y$). We are given that m($\angle A$)=m($\angle X$) and $|BC|=|YZ|$ so, by the ASA property, $\triangle ABC \simeq \triangle XYZ$.

5.   We have $\triangle PQR$ with m($\angle P$)=m($\angle R$).
      Prove: $|PQ|=|QR|$

Draw the perpendicular from Q to $\overline{PR}$ meeting it at point N. Thus m($\angle QNP$)=m($\angle QNR$) since they are right angles. We have m($\angle P$)=m($\angle R$). In addition, $|QN|=|QN|$. By ASA, we have the two congruent right triangles: $\triangle PNQ \simeq \triangle RNQ$. Since corresponding parts of congruent triangles have the same measure, $|PQ|=|QR|$.

6.   Let $\triangle ABC$ have three angles with the same measure.
      Prove: $\angle ABC$ is equilateral.

Given that the m($\angle A$)=m$\angle B$) by problem 5 it follows that $|BC|=|AC|$. Also we have m($\angle B$)=m($\angle C$). Again by problem 5 we have $|AC|=|AB|$. Thus $|AB|=|AC|=|BC|$ and $\triangle ABC$ is equilateral.

7.   Let $\triangle AHB$ and $\triangle DHC$ have $|CH|=|HB|$ and $|AH|=|HD|$.
      Prove: $\overline{AB}$ is parallel to $\overline{CD}$.

Since $|CH|=|HB|$, $|AH|=|HD|$ and m($\angle AHB$)=m(DHC) by Opposite Angle Theorem we have $\triangle AHB \simeq \triangle DHC$ by SAS. Regarding the ironing board, we now know that m($\angle HCD$)=m($\angle ABH$) because they are corresponding angles in these congruent triangles. That means that $\overline{AB}$ is parallel to $\overline{CD}$ since these alternate angles have the same measure (Problem 21, page 59). Consequently, the ironing board will remain parallel to the floor.

8.   In Thales triangles we have m($\angle D'BA$)=m($\angle DBA$),
      m($\angle D'AB$)=m($\angle DAB$). Prove: $|AD'|=|AD|$ (See figure 7.21)

We have given that m($\angle D'BA$)=m($\angle DBA$) and m($\angle D'AB$)=m($\angle DAB$). We know that $|BA|=|BA|$. By the ASA property, $\triangle ABD \simeq \triangle ABD'$. Thus, since these are corresponding parts, the land distance, $|AD'|$ equals the sea distance $|AD|$.

9.   This method of Thales is also based on the ASA congruence case. Here we have m($\angle XZY'$)=m($\angle XZY$), m($\angle ZXY'$)=m($\angle ZXY$) and $|ZX|=|ZX|$. Thus $\triangle ZXY \simeq \triangle ZXY'$. His land distance, $|XY'|$ equals the sea distance, $|XY|$.

10.   For this pliers we are given $|AE|=|EC|$ and $|BE|=|ED|$. Since $\angle AEB$ and $\angle DEC$ are opposite angles their measures are equal. By SAS we have $\triangle AEB \simeq \triangle DEC$. $m(\angle DEC)=m(\angle BAE)$ since they correspond in these congruent triangles. These alternate angles having the same measure implies that $\overline{AB}$ is parallel to $\overline{DC}$.

§2. Exercises
(Pages 195-200)

1.   In parallelogram ABCD the diagonals intersect at O.
      Prove: $|AO|=|OC|$ and $|BO|=|OD|$.

      By definition of a parallelogram, $\overline{AB}$ is parallel to $\overline{DC}$ so alternate angles $\angle OAB$ and $\angle DCO$ have the same measure. Since $\overline{BC}$ is parallel to $\overline{AD}$ we have $m(\angle ABO)=m(\angle CDO)$. By Theorem 7-1, $|CD|=|BA|$. By ASA $\triangle AOB \simeq \triangle COD$. Finally, $|AO|=|OC|$ and $|BO|=|OD|$ because they are corresponding sides in these congruent triangles.

2.   In quadrilateral MATH we are given that $|MO|=|OT|$ and $|HO|=|OA|$ where O is the intersection point of the diagonals. (See fig. 7.30) Prove: MATH is a parallelogram.

      In the triangles $\triangle HOM$ and $\triangle AOT$ we are given $|MO|=|OT|$ and $|HO|=|OA|$. Since $\angle HOM$ and $\angle AOT$ are opposite angles their measures are equal. By SAS it follows that $\triangle HOM \simeq \triangle AOT$. By corresponding parts we have $m(\angle OHM)=m(\angle OAT)$. Since these are alternate angles for $\overline{HM}$ and $\overline{AT}$, HM is parallel to $\overline{AT}$. Similarly, $m(\angle HMO)=(\angle ATO)$ so $\overline{HT}$ is parallel to $\overline{MA}$. By definition, MATH is a parallelogram.

3.   Let ABCD be a quadrilateral with $|AD|=|BC|$ and $|AB|=|DC|$. Prove: ABCD is a parallelogram

      Draw the diagonal $\overline{BD}$. In triangles $\angle ABD$ and $\angle CDB$ we are given $|AD|=|BC|$ and $|AB|=|DC|$. Since $|DB|=|DB|$ we have $\triangle ABD \simeq \triangle CDB$ by SSS. By corresponding parts $m(\angle DBA)=m(\angle CDB)$. These are alternate angles so $\overline{CD}$ is parallel to $\overline{AB}$. Also $m(\angle ADB)=m(\angle DBC)$ so $\overline{AD}$ is parallel to $\overline{BC}$. Thus ABCD is a parallelogram by definition.

4.   In quadrilateral ABCD assume that $\overline{AB}$ is parallel to $\overline{DC}$ and $|AB|=|DC|$. Prove: ABCD is a parallelogram.

Draw diagonal $\overline{BD}$. In triangles $\triangle ABD$ and $\triangle CDB$ we are given that $|AB|=|DC|$ and $\overline{AB}$ is parallel to $\overline{DC}$. By the alternate angle theorem we have m($\angle CDB$)=m($\angle ABD$). Since $|BD|=|BD|$ by SAS we get $\triangle ABD \simeq \triangle CDB$. By corresponding parts we have $|AD|=|CB|$. By Theorem 7-4 we conclude that ABCD is a parallelogram.

5. Let MATH be a rhombus and let its diagonals be drawn meeting at point O. Prove: The diagonals are perpendicular to each other and they bisect the angles of the rhombus. (See fig.7.30)

Since a rhombus is a parallelogram, by problem 1 we have that the diagonals of a parallelogram bisect each other. We consider the triangles $\triangle HOM$ and $\triangle AOM$. Since diagonal $\overline{MT}$ bisects $\overline{AH}$ we have $|HO|=|AO|$. Because MATH is a rhombus $|MA|=|MH|$. Finally $|MO|=|MO|$. By SSS $\angle HOM \simeq \triangle AOM$. By corresponding parts, we have m($\angle HOM$)=m($\angle AOM$). Because AOH is a straight angle, $\angle HOM$ and $\angle AMO$ are right angles. Thus $\overline{MT}$ is perpendicular to $\overline{AH}$. We also have m($\angle HMO$)=m($\angle AMO$). Thus $\overline{MT}$ bisects $\angle M$. Since each of the other triangles formed by a side and half of each diagonal is congruent by the same argument to the first two triangles, we conclude that each angle is bisected by the diagonal to its vertex.

6. Let ABCD be a parallelogram with a diagonal $\overline{BD}$ that bisects $\angle B$ and $\angle D$. Prove: ABCD is a rhombus.

Since $\overline{BD}$ bisects $\angle B$ and $\angle D$ we have m($\angle ABD$=m($\angle CBD$) and m($\angle ADB$)=m($\angle CDB$). With $|BD|=|BD|$, by ASA we obtain $\triangle ABD \simeq \triangle BCD$. The corresponding parts have the same measure so $|AB|=|BC|$ and $|AD|=|CD|$. Opposite sides of a parallelogram have the same measure so $|AB|=|CD|$. Thus all four sides have the same measure and ABCD is a rhombus by definition.

7. Let point O lie on the angle bisector of $\angle PQR$. (see figure 7.31) Prove: O is equidistant from the sides of the angle.

Draw OX $\perp R_{QP}$ and OY $\perp R_{QR}$. Thus m($\angle X$)=m($\angle Y$). The segment $\overline{QO}$ is common to the triangles $\triangle QOX$ and $\triangle QOY$. Because $\overline{QO}$ bisects $\angle PQR$ we have m($\angle XQO$)=m($\angle YQO$). By ASA $\triangle QOX \simeq \triangle QOY$. $|OX|=|OY|$ since they are corresponding parts of these triangles. Therefore, O is equidistant from the sides of the angle.

8. Let O be a point inside the $\angle PQR$ and equidistant from the sides of the angle. Prove: O lies on the bisector of $\angle PQR$.

From O draw perpendiculars to the sides of the angle meeting R_{QP} at X and R_{QR} at Y. Since O is equidistant from the sides $|OX|=|OY|$. Also $\angle X$ and $\angle Y$ are right angles so $m(\angle X)=m(\angle Y)$. In the right triangles $\angle QOY$ and $\angle QOX$, by the Pythagoras Theorem $|QY|^2+|OY|^2=|QO|^2$ and $|QX|^2+|OX|^2=|QO|^2$. By substitution, $|QY|^2+|OY|^2=|QX|^2+|OX|^2$. Since $|OX|=|OY|$ we get $|OX|^2=|OY|^2$. Subtracting equals we obtain $|QY|^2=|QX|^2$. Thus $|QY|=|QX|$. We also have $|QO|=|QO|$. By SSS we get $\triangle QOY \simeq \triangle QOX$. Since $\angle YQO$ and $\angle XQO$ are corresponding parts of these congruent triangles we have $m(\angle XQO)=m(\angle YQO)$. Therefore, $\overline{QO}$ bisects $\angle PQR$.

9.   (a)   Let $\triangle PQA$ be a triangle with three lines, L, M, and N, that bisect the three angles of the triangle. Let O be the point of intersection of L and M.
            Prove: Point O lies on line N.

            Let L bisect $\angle P$, M bisect $\angle Q$ and N bisect $\angle A$. Since O lies on L, Problem 7 assures that it is equidistant from $\overline{PQ}$ and $\overline{PA}$. Since it lies on M, it is equidistant from $\overline{QP}$ and $\overline{QA}$. Thus O is equidistant from $\overline{QA}$ and $\overline{PA}$ and, by Problem 8, lies on the bisector of $\angle A$ which is line N.

      (b)   Let C be an inscribed circle to the triangle $\triangle PQA$, and let M be its center.
            Prove: M is the point of intersection of the angle bisectors of $\triangle PQA$.

            Let X, Y and Z be the points of tangency on the sides of $\triangle PQA$. The radii to X, Y, and Z are perpendicular to the sides. Therefore M is equidistant from the three sides and, by Problem 8, lies on all three angle bisectors and is their point of intersection.

10.   Let $\triangle XYZ$ be equilateral with A, B, and C as the midpoints of its sides.
       Prove: $\triangle ABC$ is equilateral.

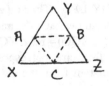

Since $\triangle XYZ$ is equilateral, $|XY|=|YZ|=|XZ|$. Because A, B, and C are midpoints of the sides $|XA|=|AY|=|YB|=|BZ|=|ZC|=|CX|$. By the Corollary to Theorem 5-3 (page 141), $m(\angle X)=m(\angle Y)=m(\angle Z)$. Therefore, by SAS $\triangle XCA \simeq \triangle YBA \simeq \triangle ZBC$. $|AC|=|CB|=|BA|$ by the corresponding parts theorem for congruence. Thus ΔABC is equilateral by definition.

11.   Let ΔABC be an equilateral triangle with line L bisecting ∠B.
       Prove: Line L is the perpendicular bisector of $\overline{AC}$.

Name the intersection of L and $\overline{AC}$ point D. Since ΔABC is equilateral, |AB| = |BC|. L bisects ∠B so m(∠ABD)=m(∠CBD). Using the fact that |BD| = |BD| we get ΔABD ≃ ΔCBD by SAS. As corresponding parts, |AD| = |DC| so L bisects AC. Again, as corresponding parts, m(∠BDA)=m(∠BDC). They form a linear pair so m(∠BDA)=m(∠BDC)=90°. Thus L is the perpendicular bisector of $\overline{AC}$.

12.   In figure 7.33 let POR and QOS be straight line segments and let m(∠1)=m(∠2) and m(∠3)=m(∠4).
       Prove: O is the midpoint of QS.

We are given that m(∠1)=m(∠2) and m(∠3)=m(∠4). Since |PR| = |PR| we have ΔPQR ≃ ΔPSR by ASA. By corresponding parts, |PQ| = |PS|. Since m(∠1)=m(∠2), and |PO| = |PO| we get ΔPQO ≃ ΔSPO. Therefore, |QO| = |SO| by corresponding parts. We can conclude that O is the midpoint of $\overline{QS}$.

13.   In figure 7.34, ECBF is a parallelogram and m(∠A)=m(∠D). Also, EFA and DCB are straight line segments.
       Prove: ΔEDC ≃ ΔFAB.

Since ECBF is a parallelogram, m(∠ECB)=m(∠EFB) because opposite angles have equal measure. Because they form linear pairs of angles, m(∠DCE)+m(∠ECB)=180° and m(∠BFA)+m(∠EFB)=180°. Therefore, m(∠DCE)+m(∠ECB)=m(∠BFA)+m(∠EFB). Subtracting equals from each side gives us m(∠DCE)=m(∠BFA). Opposite sides of a parallelogram are equal so |EC| = |BF|. Given that m(∠A)=m(∠D), by ASA we have ΔEDC ≃ ΔBFA.

14.   In the construction of an angle equal to a given angle (p. 199), we have |OS| = |PQ|, |OT| = |PR| and |ST| = |QR|. By SSS, ΔOST ≃ ΔPQR. Thus m(∠O)=m(∠P).

§ 3. Exercises
(Pages 206-209)

1.   In figure 7.47, we are given $|AC|=4$, $|CB|=4$, $|AB|=4\sqrt{2}$, $|YZ|=4$, $m(\angle Z)=90°$ and $m(\angle Y)=45°$. Prove: $\triangle ABC \simeq \triangle XYZ$.

In $\triangle XYZ$ we have a 45° right triangle with $|YZ|=4$ so $|ZX|=4$ and $|XY|=4\sqrt{2}$. Thus $|BC|=|YZ|$, $|AC|=|XZ|$ and $|AB|=|XY|$. Therefore, $\triangle ABC \simeq \triangle YXZ$ by SSS.

2.   In $\triangle TOR$ draw $\overline{ON}$ perpendicular to $\overline{RT}$. $\triangle TON$ is a 30-60-90 triangle. $|OT|=8$ so $|ON|=4$ and $|NT|=4\sqrt{3}$ and $|TR|=8\sqrt{3}$. The area of $\triangle ROT$ is $(8\sqrt{3})(4)/2=16\sqrt{3}$.

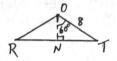

3.

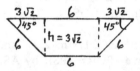

In the diagram above note that $\angle A=45°$ and the perpendiculars from B and C to AD are the height of the trapezoid, h.
$h=6\sqrt{2}=(6\sqrt{2})(\sqrt{2}/\sqrt{2})=3\sqrt{2}$. The long side of the trapezoid has length $3\sqrt{2}+6+3\sqrt{2}=6+6\sqrt{2}$.
$A=(1/2)(3\sqrt{2})(6+(6+6\sqrt{2})=(1/2)(3\sqrt{2})(12+6\sqrt{2})=(1/2)(36\sqrt{2}+36)$
$=18\sqrt{2}+18$.

4.   In this trapezoid the altitude is $h=3\sqrt{2}/\sqrt{2}=3$. The long side's length is $3+6+3=12$. Therefore the area is $(1/2)(3)(6+12)=(1/2)(3)(18)=27$.

5.   In figure 7.51, note the 30-60-90 triangle in which the longer leg is 3 m. The ratio of legs is $\sqrt{3}:1$ so the shorter leg is $3/\sqrt{3}=\sqrt{3}$ m. Build the walls $\sqrt{3}$ m from the sides.

6.

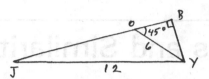

In the 45-45-90 triangle $\triangle$BOY
$|BO|=|BY|=6\sqrt{2}=(6\sqrt{2})(\sqrt{2}/\sqrt{2})=3\sqrt{2}.$
$|JB|^2=12^2 - (3\sqrt{2})^2=144 - 18=126$ so $|JB|=\sqrt{126}=3\sqrt{14}.$
Area of $\triangle$BOY$=(1/2)(3\sqrt{2})(3\sqrt{2})=9$
Area of $\triangle$BJY$=(1/2)(3\sqrt{14})(3\sqrt{2})=(9\sqrt{28})/2=(18\sqrt{7})/2=9\sqrt{7}.$
Area of $\triangle$JOY$=$Area of $\triangle$BJY - Area of $\triangle$BOY $= 9\sqrt{7}$-9

7.   In figure 7.53 we are given that $\triangle$ABC is equilateral and $\angle$BAD is a
       right angle. Prove: $|BC|=|CD|$.

       Since $\triangle$ABC is equilateral, by the Isosceles Triangle Theorem we get
       m($\angle$B)=60°. Thus $\triangle$ABD is a 30-60-90 right triangle where
       $|AB|=(1/2)|BD|$ or$|BD|=2|AB|$. By the SEG Property,
       $|BD|=|BC|+|CD|$. By substitution we get $2|AB|=|BC|+|CD|$.
       Since $|AB|=|BC|$ in the equilateral triangle, we have
       $2|BC|=|BC|+|CD|$. By subtraction we conclude $|BC|=|CD$.

8.   In figure 7.54 we are given $\overline{UI}$ parallel to $\overline{QE}$ and $\overline{QU}$ parallel to $\overline{EI}$.
       By definition, QUIE is a parallelogram. Thus $|QU|=|EI|$ and
       $|QE|=|UI|$. The area of QUIE is bh=6(10)=60 sq. cm.
       $|QE|^2=10^2+5^2=100+25=125$. $|UI|=|QE|=\sqrt{125}=5\sqrt{5}$ cm.

9.   In figure 7.55 we have $\overline{XM}$ parallel to $\overline{SA}$, $\overline{MA}$ parallel to $\overline{XS}$ and $\overline{MO}$
       perpendicular to $\overline{OS}$.

       By definition, XMAS is a parallelogram. Hence, $|XM|=|SA|=7$. the
       area of XMAS $=$bh$=(7)(5)=35$ sq. units

CHAPTER 8

Dilations and Similarities

§ 1. Exercises
(Pages 214-216)

1.

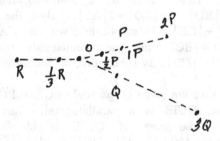

2. $D_r(P) = P$ when $r = 1$.

3. Points for this image set have 3 as a subscript.

4. Points for this image set have 4 as a subscript and the triangle is a dashed line.

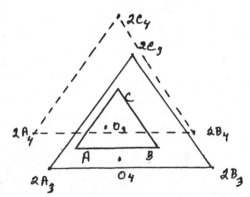

5.

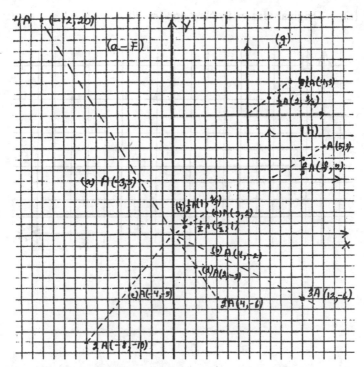

6.

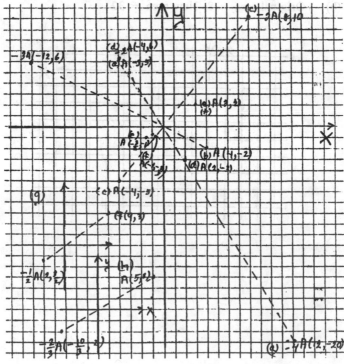

7.   (a)   Original area = 9, New area = 36

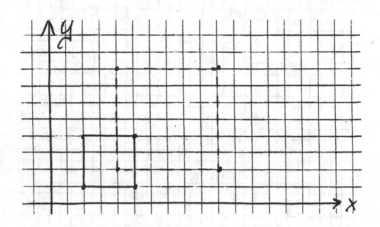

(b)   Original area = 8, New area = 32

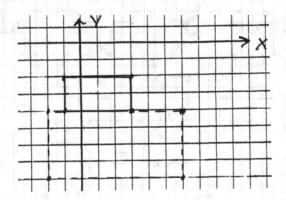

8.   (a)   Original area = 4, New area = 9

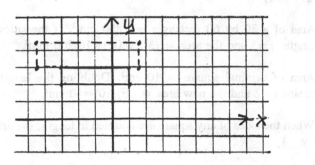

(b)   Original area = 10, New area = 22.5

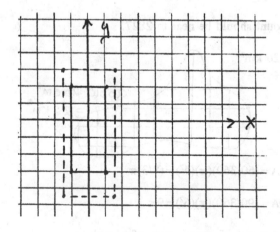

§ 2. Exercises
(Pages 227-231)

1.   Area of a 30 by 60 rectangle = 1800 mm². If the sides are tripled in length, r=3 and the area = $(3^2)(1800)=16,200$ mm².

2.   Area of original square = 10 cm². Doubling the lengths of the sides means r=2 and the new area = $(2^2)(10)=40$ cm².

3.   When the sides of any square are doubled in length, the area is multiplied by _4_.

4.   If the radius of a disk is doubled, its area is multiplied by _4_.

5.   Area of original table = $1.5^2=2.25$ m².
      New area = $4.5=r^2(2.25)$ so $r^2=2$ and $r=\sqrt{2}$. Each side should be $(\sqrt{2})(1.5)$ m.

6.   If you wish to double the area of a square, you must multiply the sides by _$\sqrt{2}$_.

7.   The smaller lot is one-fourth the size of the larger lot. A fair price for the smaller lot would be (1/4)(2000)=\$500.

8.   Your count should be near to 2827.

9.   About 26 km².

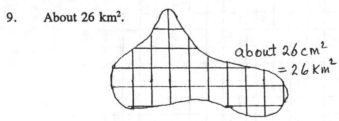

about 26 cm²
= 26 km²

10.   (a) $A=(60/360)(\pi)(4^2)=(8/3)\pi$

      (b) $A=(90/360)(\pi)(6^2)=9\pi$

      (c)   A=area of sector - area of triangle
            $= (1/6)(\pi)(36) - (6^2/4)(\sqrt{3})=6\pi - 9\sqrt{3}$.

      (d) $A=(1/4)(\pi)(25) - (1/2)(5)(5)=(25/4)(\pi) - 25/2$

11. (a) $(1/4)(\pi)(1^2)=\pi/4$

    (b) $\pi/3$

    (c) $\pi/12$

    (d) $(x/360)\pi$

12. (a) $[(\theta_2-\theta_1)/360]\pi r^2$

    (b) $\pi r_2^2 - \pi r_1^2 = \pi(r_2^2 - r_1^2)$

    (c) $[(\theta_2 - \theta_1)/360][\pi(r_2^2 - r_1^2)]$

13. In figure 8.32 draw $\overline{PQ}$. Because of the symmetry in the figure, we can compute the area of S (one-half of the total area) as the area of the sector in the left circle minus the area of the triangle OPQ.
    Area $= (1/4)(\pi)(36)-(1/2)(6)(6)=9\pi-18$. Thus the total area $=18\pi-36$.

14. The area of the shaded region is the area of the square minus the areas of S_1, S_2 and S_3.

    Area of square $= 64$
    Area of S_1 = area of S_2 = $(1/4)\pi(16)=4\pi$
    Area of S_3 = $4^2=16$
    Area of shaded region = $64 - (4\pi+4\pi+16)=48-8\pi$

15. $\pi(5^2-3^2)=\pi(25-9)=16\pi$.

16. In figure 8.35, $m(\angle BOA)=360/8=45°$.
    $|BO| = |OA| =r$ and $|BX| =r\sqrt{2}=(r\sqrt{2})/2$.
    Area of $\triangle BOA$ = $(1/2)|OA||BX| =(1/2)(r)(r\sqrt{2}/2)=(r^2\sqrt{2})/2$
    The 8 triangles of the octagon are congruent by SAS thus the area of the octagon = $8[(r^2\sqrt{2})/4]=2r^2\sqrt{2}$.

§ 3. Exercises
(Pages 234-235)

1. Area of the dilated square $= (3^2)(25) = 225$.
Perimeter of dilated square $= (3)(20) = 60$

2. Let T have sides 2, 4, 5. Perimeter of $T = 2+4+5 = 11$.
The dilation of the short side, 2, equals 7 so $2d = 7$ and $d = 7/2$.
Thus, the perimeter of the dilated triangle $= (7/2)(11) = 77/2 = 38\ 1/2$.

3. If r is the dilation factor, $rP_1 = P_2$. $r = 10/16$ (or $16/10$).
$r^2 A_1 = A_2$ so $A_2/A_1 = (10/16)^2 = 100/256$.

4. $A_2/A_1 = a^2/b^2$.

§ 4. Exercises
(pages 240-243)

1. (a) For the inscribed square, the side of the triangle is approximately 7 cm and the height is about 3.5 cm. The area of the triangle is $(1/2)(7)(3.5) = 12.25$. The area of the square is $(4)(12.25) = 49\ \text{cm}^2$. For the circle the area is $25\pi = 78.5\ \text{cm}^2$. The perimeter of the square is 28 cm and the circumference of the circle is 31.4 cm.

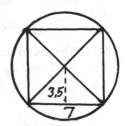

    (b-e) Follow similar procedures.

2. Perform the measurements.

3. (a) $A = 24\pi = \pi r^2$ so $r^2 = 24$ and $r = \sqrt{24} = 2\sqrt{6}$.
Thus $C = (2)(\pi)(2\sqrt{6}) = 4\pi\sqrt{6}$ cm.

    (b) $A = 36\pi = \pi r^2$ so $r^2 = 36$ and $r = 6$.
Thus $C = (2)(\pi)(6) = 12\pi$ cm.

4. (a) $C = 15\pi = 2\pi r$ so $r = 15/2$.
Thus $A = \pi(15/2)^2 = (225/4)\pi\ \text{cm}^2$.

    (b) $C = 9\pi = 2\pi r$ so $r = 9/2$.
Thus $A = \pi(9/2)^2 = (81/4)\pi$.

(c) $C=20\pi$ so $r=10$. $A=100\pi$ cm^2.

5.   The surface area of a cylinder is the same as a rectangle with length $C=2\pi(3)=6\pi$ m and height $=5$ m. $A=5(6\pi)=30\pi$ m^2.

6.   Total area $=$ half the circumference times the height. $C=2\pi(4)=8\pi$. $(1/2)C=4\pi$. Thus $A=4\pi(11)=44\pi$ m^2.

7.   Area of roof $=10$ times half the circumference$=10((1/2)(9\pi)=45\pi$. Area of the sides $=4$ rectangular areas $+2$ semicircles. $A=(2)(8)(9)+(2)(8)(10)+2(1/2)(\pi)(9/2)^2=144+160+(81/4)\pi$ $=304+(81/4)\pi=304+(20\ 1/4)\pi$ m^2. Thus the total area $=304+(65\ 1/4)\pi$ m^2.

8.   Area of the wasted space $=$ area of square - area of four circles. $A=(12)^2$ $-(4)(\pi)(3)^2=24-36\pi$ cm^2.

9.   If the radius is doubled, the circumference is doubled.

10.   Let $S=$ length of the strip. $S=(2)(12)+(2)(6)+(4)(1/4)(6\pi)=24+12+6\pi=36+6\pi$ cm.

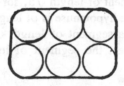

11.   Let $S=$ speed of the bicycle $=15$ km/hr. Let $N=$ number of revolutions/hr. Circumference of the wheel$=1.2\pi$ m$=(.0001)(1.2\pi)$ km$=.0012\pi$ km. $N=S/C=(15)/(0.0012\pi)=(12500)/\pi$ revolutions/hr. This is approximately 3979 revolutions/hr.

12. $P=2\pi r$ so $r=P/(2\pi)$. $A=\pi r^2=\pi(P/2\pi)^2=\pi(P^2/4\pi^2)=P^2/(4\pi)$.

13.   (a) $V=$cross section area times the average length. $V=(\pi b^2)(2\pi a)=2a\pi^2b^2$

       (b)   SA$=$circumference of the torus times its length. $SA=(2\pi b)(2\pi a)=4ab\pi^2$.

14.   Let $C'=C+1$. Then $2\pi r'=2\pi r+1$ so $r'=r+1/(2\pi)$. The increase in the radius is $1/(2\pi)$ m. In other words, the increase in the radius is independent of the original r.

Additional Exercises on Circles
(Pages 243-244)

1. In the figure below, draw AC perpendicular to OB. ΔAOC is a 45-45-90
    right triangle with hypotenuse x. $|OC| = |AC| = (x\sqrt{2})/2$. Thus $|CB| = x-$
    $(x\sqrt{2})/2$ by Pythagoras.

    $$|AB|^2 = [(x\sqrt{2})/2]^2 + [x-(x\sqrt{2})/2]^2$$
    $$= [x^2(2/4)] + [x^2(1-\sqrt{2}+(1/2)]$$
    $$= x^2[(1/2)+1+(1/2)-\sqrt{2})$$
    $$= x^2(2-\sqrt{2})$$

    Therefore, $|AB| = x\sqrt{2-\sqrt{2}}$.

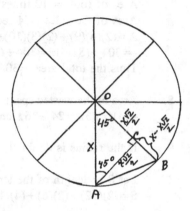

2. To construct a segment of length $\sqrt{5}$, form a right triangle with legs of
    lengths 1 and 2. The hypotenuse is of length $\sqrt{5}$. To construct the length
    $(1/2)(\sqrt{5}-1)$, mark a segment of length $\sqrt{5}$ on a line and subtract the
    length 1. Then bisect the segment of length $(\sqrt{5}-1)$. One half is the
    required length.

3. In the figure below draw $\overline{OB}$, $\overline{CO'}$, $\overline{OO'}$ and $\overline{AO'}$ parallel to $\overline{BC}$.
    $|AO| = 25-9 = 16$. $|OO'| = 25+9 = 34$. Since ABCO' is a rectangle
    $|BC| = |AO'|$.
    $|BC|^2 = 34^2 - 16^2$
    $\qquad = 1156 - 256 = 900$ so
    $|BC| = 30$.

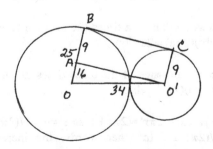

4.   (a)

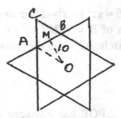

We are given that $|OM|=10$. $\triangle AOM$ is a 30-60-90 right triangle. The ratio of the legs is $1/\sqrt{3}$ so $1/\sqrt{3}=x/10$ so $x=10/\sqrt{3}$. Thus $|AB|=20/\sqrt{3}$. Therefore the perimeter,
$P = 120/\sqrt{3}$.
The area of the hexagon $= 6(1/2)(20/\sqrt{3})(10)=600/\sqrt{3}$.

(b) $\triangle ABC \approx \triangle ABO$ by ASA since
$m(\angle CAB)=m(\angle CBA)=m(\angle BAO)=m(\angle ABO)=60°$ and AB is a common side. $|AO|=2|AM|=20/\sqrt{3}$ so $|AB|=|AO|=20/\sqrt{3}$. The perimeter of the star is equal to $P=(12)(20/\sqrt{3})=240/\sqrt{3}$.
The area twice the area of the hexagon $= 1200/\sqrt{3}$.

5. $C=16\pi=2\pi r$ so $r=8$. One side of the hexagon $= r$. The perimeter $= 6r=48$.

6. $C=80\pi=2\pi r$ so $r=40$. In the 45-45-90 triangle, $s=40\sqrt{2}$.

7.   The number of revolutions per minute, $N=350/\text{minute}$. The circumference of the wheel is $C=\pi d=20\pi$ cm. The ground speed, $S=350(20\pi)=7000\pi$ cm/minute.

8. $C=10\pi=2\pi r$ so $r=5$. One side of the square is of length 10 so the perimeter $= 40$.

9.   Given the length of arc, $L=4\pi$, and $C=2\pi r=2\pi(6)=12\pi$ we have $(4\pi)/(12\pi)=x/360$. Therefore, $x=360/3=120°$.

10. For the 72° arc the length, $L=4\pi$. Thus $C/(4\pi)=360/72$ so $C=20\pi$. Therefore, $20\pi=2\pi r$ so $r=10$.

11. $C=10\pi=2\pi r$ so $r=5$. $A=\pi(5^2)=25\pi$.

12. The diagonal of the inscribed rectangle is the diameter of the circle. $d^2=8^2+12^2=64+144=208$ so $d=\sqrt{208}=4\sqrt{13}$. Thus $r=2\sqrt{13}$. $A=\pi(2\sqrt{13})^2=52\pi$.

13.   For circle A, area$=16=\pi r^2$ so $r^2=16/\pi$ and $r=4/\sqrt{\pi}$. The radius of circle B $=2/\sqrt{\pi}$. Area of B $=\pi(2/\sqrt{\pi})^2=4$. (By a dilation, circle B is dilated by 1/2 from circle A. Thus the area of circle B $=$ 1/4 area of circle A.)

14.   Since inscribed angle $\angle$ PQR has measure 150°, the arc PSR has twice the measure of $\angle$ PQR, 300°. Thus the measure of arc PQR is 360-300$=60°$.
      The length of arc PQR, L$=(60/360)(2\pi)(10)=10\pi/3$.

§ 5. Exercises
(Pages 254-260)

1. $D_r(T)=T'$ so $r=a'/a$. Let b, a side of ΔT, have length 7.

      (a) $a'/a=4/2=b'/7$ so $b'=28/2=14$ cm.

      (b) $a'/a=8/3=b'/7$ so $b'=56/3$ cm.

      (c) $a'/a=5/1=b'/7$ so $b'=35$ cm.

2.   (a)   In figure 8.63(a), $r=|$CB'$|/|$CB$|=2/9$. The dilation at C with ratio 2/9 takes $\overline{AB}$ to $\overline{AB}$'. $D_{2/9,c}(\overline{AB})=\overline{A'B'}$.

      (b)   In figure 8.63(b), $r=|\overline{CA}'|/|\overline{CA}|=3/1$. $D_{3,c}(\overline{AB})=\overline{A'B'}$

3.   (a)   In figure 8.64(a), $x/4=4/5$ so $x=16/5$ by Theorem 8-8.

      (b)   In figure 8.64(b), $x/3=11/4$ so $x=33/4$ by Theorem 8-8.

4.   In figure 8.65, we are given that m($\angle$ ADE)$=$m($\angle$C).
      Prove: ΔABC similar to ΔADE.

      We are given m($\angle$ADE)$=$m($\angle$C) and we also have m($\angle$A)$=$m($\angle$A).
      Therefore ΔABC is similar to ΔAED by the AA similarity property.
      By this similarity $\overline{AB}/\overline{AE}=\overline{BC}/\overline{ED}$. Let $|$AB$|=x$.
      Thus, $x/3=12/7$ and $x=36/7$.

5. By the AA similarity Property the two triangles below are similar. Therefore, $8/2 = x/1$ so $x = 4$ m.

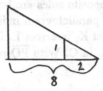

6. In figure 8.66, let $x =$ tree height. The two triangles are similar by AA. Therefore, $x/2 = 20/2$ so $x = 20$ m.

7. Draw $\overline{AC}$ perpendicular to $\overline{AB}$ with C far enough north to have a line of sight to B that is over the land. At some intermediate point on $\overline{AC}$, draw a perpendicular to $\overline{AC}$ at P meeting $\overline{CB}$ at Q. Then $\triangle ABC$ is similar to $\triangle PQC$. The proportion $\overline{CA}/\overline{CP} = \overline{AB}/\overline{PQ}$ allows one to obtain $|AB|$ only on the ground.

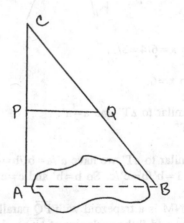

8. $A_2/A_1 = 4/9 = r^2$. Thus $r = 2/3$. Therefore $P_2/P_1 = 2/3$.

9. $r = 10/8$. Let x,y be the other sides.
    Thus $10/8 = x/5$ and $x = 50/8 = 25/4$,
    and $10/8 = y/7$ and $y = 70/8 = 35/4$.

10. A sufficient condition for two rectangles to be similar is that the adjacent sides of one rectangle have the same ratio as the corresponding sides of the other rectangle.
    There may be other conditions.

11. $r = 18/12 = 3/2$. Therefore, $w_2/w_1 = 3/2$ and $P_2/P_1 = 3/2$.

12. $r = 10/5 = 2/1$. Thus $2/1 = a/7 = b/8$ so $a = 14$ and $b = 16$.

13.   In figure 8.68, let the intersection of L_1 and K' be point A. Let K_1 be the line through A, parallel to K. Then K_1, L_1, L_2 and K determine a parallelogram so the opposite sides are both of length x. Similarly, with L_2 and L_3 another parallelogram indicates that the opposite sides are both of length y. Let K_1 intersect L_2 in P and L_3 in B. Let K' intersect L_2 in Q and L_3 in C. With $\overline{PQ}$ parallel to $\overline{BC}$ in $\triangle ABC$, by Theorem 8-8, x/y=u/v.

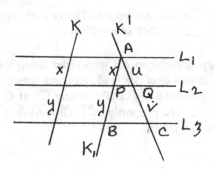

14.   (a) x/3=2/4 so x=6/4=3/2.

       (b) x/3=2/1 so x=6.

15.   Given $\triangle T$ is similar to $\triangle T'$ and a=a'.
       Prove: $\triangle T \simeq \triangle T'$

       Since $\triangle T$ is similar to $\triangle T'$ we have a'/a=b'/b=c'/c. We are given that a=a' so a'/a=1=b'/b=c'/c. So b=b' and c=c'. Therefore, $\triangle T \simeq \triangle T'$.

16.   Given that PQNM is a trapezoid with $\overline{PQ}$ parallel to $\overline{MN}$. Let R be the midpoint of $\overline{PM}$ and S the midpoint of $\overline{NQ}$ and let b be the length of $\overline{RS}$.
       Prove: (a) The area of the trapezoid = bh were h is the distance between $\overline{MN}$ and $\overline{PQ}$. (b) $b=(b_1+b_2)/2$

       As in figure 8.70, draw a segment perpendicular to $\overline{PQ}$ at A, through R and meeting the line $\overleftrightarrow{MN}$ in T. Draw another segment perpendicular to $\overline{PQ}$ at B, through S and meeting line $\overleftrightarrow{MN}$ in U.
       $m(\angle TRM)=m(PRA)$ by the opposite angles theorem. Since R is the midpoint of MP, $|MR|=|PR|$ and $|RA|=|PT|$. By SAS $\triangle MTR \simeq \triangle RAP$. By the same argument $\triangle NUS \simeq \triangle SQB$. Consequently, the area of the trapezoid = area of the rectangle ABUT. hence A=bh. For part (b) we note $A=bh=(1/2)h(b_1+b_2)$. Dividing by H we obtain $b=(b_1+b_2)/2$.

17.   (a)   △ABC is similar to △XYZ. Let |AN| and |XP| be their heights.
            Prove: |AN|/|XP|=|AB|/|XY|.

            Since △ABC is similar to △XYZ we have m($\angle$B)=m($\angle$Y).
            Using the same chord, m($\angle$BNA)=m($\angle$YPX)=90°. Therefore
            △BNA is similar to △YPX by AA. Thus
            |AN|/|XP|=|AB|/|XY|.

      (b)   In the figure below $\overline{BC}$ is parallel to $\overline{AD}$ so m($\angle$1)=m($\angle$2) and
            m($\angle$3)=m($\angle$4) by alternate angles theorem. Thus △BOC is
            similar to △DOA. The corresponding parts are in the same ratio
            and |BC|/|AD|=h_1/h_2. Let h_2=x and h_1=12-x. Then
            9/15=3/5=(12-x)/x. Thus 3x=60-5x so 8x=60 and
            x=60/8=15/2. So point O is 15/2 cm from side $\overline{AD}$.

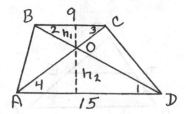

18.   In figure 8.72 we are given right △ABC with $\overline{CN}$ perpendicular to $\overline{AB}$.
      Prove: (i) b^2=cd, (ii) a^2=ce and (iii) h^2=de.

      Since m($\angle$ACN)=m($\angle$B) and m($\angle$NCB)=m($\angle$A) by AA we have
      △ABC similar to △ACN and △CBN.
      (i) Using △ABC and △ACN, c/b=b/d so b^2=cd.
      (ii) Using △ABC and △CBN, c/a=a/e so a^2=ce.
      (iii)Using △ACN and △CBN, d/h=h/e so h^2=de.

19.   Let e=6 and h=8. Find a, b, c, and P.
      8^2=6d so d=64/6=32/3 and c=32/3 + 6=50/3.
      b^2=(50/3)(32/3)=1600/9 and b=40/3
      a^2=(50/3)(6)=100 so a=10.
      P=10+(40/3)+(50/3)=120/3=40

20.   From problem 19, let |BD|=h=8, |DC|=e=4, |AD|=d, |AB|=b,
      |AC|=c and |BC|=a. Find d, a and b.
      8^2=4d so d=64/4=16 and c=16+4=20
      a^2=(20)(4)=80 so a=$\sqrt{80}$=4$\sqrt{5}$.
      b^2=(20)(16)=320 so b=$\sqrt{320}$=8$\sqrt{5}$. Therefore, |AD|=16,
      |BC|=4$\sqrt{5}$ and |AB|=8$\sqrt{5}$.

21.   In figure 8.74, draw ΔXMN. Since ∠X is inscribed in a semi-circle,
      m(∠X) = 90°. Thus ΔXMN is a right triangle and by problem 18 we
      have $h^2=(1)(4)=4$ so $h=2$ m.

22.   Given ΔABC with P, Q and R midpoints of the 3 sides.
      Prove: ΔPQR is similar to ΔABC. (See fig. 8.75)

      Since P, Q and R are midpoints the following dilations can be used to
      prove this result. Recall $D_{2,A}$ means the dilation with ratio 2 from point
      A.
      $D_{2,A}(\triangle APR)=\triangle ABC$ which means $BC/PR=2$ by theorem 8-3.
      $D_{2,B}(\triangle PBQ)=\triangle ABC$ so $\overline{AC}/\overline{PQ}=2$.
      $D_{2,C}(\triangle RQC)=\triangle ABC$ so $\overline{AB}/\overline{RQ}=2$.
      Therefore, $\overline{BC}/\overline{PR}=\overline{AC}/\overline{PQ}=\overline{AB}/\overline{RQ}$ and ΔPQR is similar to ΔABC by
      SSS similarity.

23.

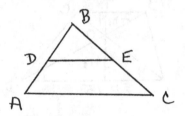

      Given ΔABC with $\overline{DE}$ joining the midpoints on $\overline{AB}$ and $\overline{BC}$.
      Prove: (1) $\overline{DE}$ is parallel to $\overline{AC}$; and, (2) $|DE|=(1/2)|AC|$.

      Since D and E are midpoints, $|AB|/|BD|=|BC|/|BE|=2$. Thus the
      dilation $D_{2,B}(\triangle DBE)=(\triangle ABC)$. By definition, ΔDBE is similar to
      ΔABC. Since corresponding sides are proportional and corresponding
      angles have equal measures in similar triangles, we have
      (1) m(∠A)=m(∠BDE) so $\overline{DE}$ is parallel to $\overline{AC}$ because
      corresponding angles have the same measure. And
      (2) $|AB|/|BD|=|AC|/|DE|=2$ which means $|DE|=(1/2)|AC|$.

24.   In quadrilateral MEAN, P, Q, R and S are midpoints of the four sides.
      Prove: PQRS is a parallelogram. See figure 8.76.

      Draw the diagonal $\overline{MA}$ forming ΔNMA and ΔMAE. In each of these
      triangles a line segment joins the midpoints of two sides. By problem
      23 these segments are parallel to $\overline{MA}$ and parallel to each other. Thus
      $\overline{SR}$ is parallel to $\overline{PQ}$. Draw $\overline{NE}$ and by the same argument $\overline{RQ}$ is
      parallel to $\overline{SP}$. Thus PQRS is a parallelogram by definition.

25.   Prove the Pythagorean Theorem

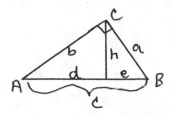

Let ▵ABC be a right triangle with the right angle at C. Draw the altitude to the hypotenuse and let its length be h. By Problem 18, $a^2 = ce$ and $b^2 = cd$. Adding these equations we obtain $a^2 + b^2 = ce + cd = c(e + d) = c(c)$, Thus $a^2 + b^2 = c^2$.

CHAPTER 9

Volumes

§ 1. Exercises
(Pages 268-270)

1. Since the area formula for a triangle is (1/2)bh, the areas are equal because they have the same base and height. The height remains the same because the distance between the parallel lines is constant. Another observation is that the area remains the same because a shearing transformation does not change the area by Theorem 5-2.

2. (a) $V=(1)(2)(3)=6 \text{ m}^3$.

    (b) $V=(a)(2a)(3/2)a=3a^3$.

    (c) $V=(8)(6)(10)=480 \text{ cm}^3$. (The bases are rectangles.)

3. (a) $SA=2(3)(2)+2(1)(3)+2(1)(2)=12+6+4=22 \text{ m}^2$.

    (b) $SA=2(a)(3/2)a+2(2a)(3/2)a+2(a)(2a)=13a^2$.

    (c)   Note that the front and back are parallelograms with area $A=(8)(10)=80$. The top and bottom are rectangles whose area is $A=(8)(6)=48$. The right and left sides are rectangles with area $A=(6)(20/\sqrt{3})=6(20\sqrt{3}/3)=40\sqrt{3}$ where the height of this rectangle is the hypotenuse of a 30-60-90 triangle with the long side = 10. Thus, $SA=2(80)+2(48)+2(40\sqrt{3})=256+80\sqrt{3} \text{ cm}^2$.

If the edge is multiplied by r, $V' = (re)^3 = (r^3)(e^3)$ so $V' = r^3V$. $SA' = 6(re)^2 = r^2(6e^2$ so $SA' = r^2SA$.

5. (a) $V = (\pi)(4^2)(25) = 400\pi$ cm^3.

    (b) $V = (\pi)(r^2)(2r) = 2\pi r^3$.

    (c) $V = (\pi)(5^2)(8) = 200\pi$ cm^3.

6. $V' = (\pi)(2r)^2(.5h) = (\pi)(4r^2)(.5h) = (2)(\pi r^2 h)$ so $V' = 2V$.

7. The surface area, SA, includes the sides, base, and top. $SA = 6(4)(10) + 2(6)(4\sqrt{3}) = 240 + 48\sqrt{3}$ cm^2 where each hexagonal base has area equal to 6 equilateral triangles with area $= (1/2)(4)(2\sqrt{3})$.

    $V = Bh = (6)(4\sqrt{3})(10) = 240\sqrt{3}$ cm^3.

8. Treat the end of the building as the base which has a surface area consisting of a 3 m by 15 m rectangle and a triangle with base 15 m and height 2 m. The height of the prism is the length of the building, 45. Then, $V = [(3)(15) + (1/2)(15)(2)][45] = [45 + 15][45] = (60)(45) = 2700$m^3.

9. A ml is 1 cm^3 so we consider a cylinder with radius 3 cm and height h that has a volume equal to 10 cm^3. $10 = \pi(3^2)(h) = 9\pi h$ so $h = 10/9\pi$ cm. This is approximately 0.35 cm.

§ 2. Exercises
(Page 274)

1. The dilated volume would be r^4 times the original volume.

2. The dilated volume would be r^n times the original volume.

§ 3. Exercises
(Pages 280, 281)

1. (a) $V=(1/3)Bh=(1/3)(\pi 4^2)(10)=160\pi/3$ units^3.

   (b) $V=(1/3)(\pi 1^2)(5\ 1/2)=11\pi/6$ units^3.

   (c) The slant height = 4.
        The diagonal of the base = $4\sqrt{2}$.
        Half of the diagonal is the base of
        the triangle with the height of the
        pyramid as the other leg. By Pythagoras
        $h^2+(2\sqrt{2})^2=4^2$ so $h^2=16-8=8$, $h=2\sqrt{2}$.
        $V=(1/3)(4)(4)(2\sqrt{2})=32\sqrt{2}/3$ units^3.

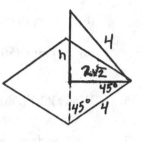

   (d) $h^2+3^2=6^2$ so $h^2=36-9=27$ and $h=3\sqrt{3}$.
        $V=(1/3)(\pi 3^2)(3\sqrt{3})=9\pi\sqrt{3}$ cm^3.

   (e) The base of the pyramid is an isosceles triangle with sides 10, 10
        and 8. For the altitude to the side of length 8, $a^2+4^2=10^2$ by
        Pythagoras. Thus $a^2=100-16=84$ and $a=2\sqrt{21}$.
        $V=[1/3][(1/2)(8)(2\sqrt{21})][12]=32\sqrt{21}$ units^3.

2. If the radius of the cone is doubled $V'=(1/3)(\pi(2r)^2)(h)=4(1/3)(\pi r^2)(h)$.
   Thus, $V'=4V$.

3. $V=(1/3)Bh=(1/3)(220)(146.59)=10,749.93$ m^3.
   Round the height to 147. Then, $N=(1/3)(220)(147)=10,780$.

4. (a) Area of the base = $(6)(1/2)(8)(4\sqrt{3})=96\sqrt{3}$.
        $V=(1/3)(96\sqrt{3})(10)=320\sqrt{3}$ units^3.

   (b) In $\triangle AOP$, $|AP|^2=10^2+8^2=164$ so $|AP|=\sqrt{164}=2\sqrt{41}$.
        One slant surface such as $\triangle PAB$ has area $A=(1/2)(8)(h)$ where
        $h^2+4^2=(2\sqrt{41})^2$ so $h^2=164-16=148$ and $h=2\sqrt{37}$. Thus, the area
        of $\triangle PAB$ = $(1/2)(8)(2\sqrt{37})=8\sqrt{37}$.
        Lateral surface area = $6(8\sqrt{37})=48\sqrt{37}$.
        Total surface area = $48\sqrt{37}+96\sqrt{3}$ units^2.

5. The square base has sides of length s and diagonals of length $s\sqrt{2}$. As in problem 1(c) consider the right triangle with its base half of the diagonal of the square and its other leg the height of the pyramid. Thus, $h^2+(s\sqrt{2}/2)^2=s^2$ so $h^2=s^2-(2s^2/4)=(4s^2-2s^2)/4=(2s^2)/4$ and $h=s\sqrt{2}/2$. $V=(1/3)(s^2)(s\sqrt{2}/2)=s^3\sqrt{2}/6$. Thus $V=s^3\sqrt{2}/6$.

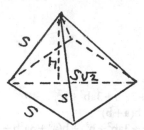

6. Let △AYC be the triangular base of this pyramid with B as its top vertex. Let △DXE be a triangular cross section parallel to △AYC. Assume △ABC is the front face of the pyramid. Since $\overline{DE}$ is parallel to $\overline{AC}$, we have parallel angles equal in measure so m(∠BDE)=m(∠BAC) and m(∠BED)=m(∠BCA). Therefore, △BDE is similar to △BAC. By the same kind of argument, △BXD is similar to △BYA and △BXE is similar to △BYC. Since corresponding sides of similar triangles are in proportion, we have

(1) △BDE and △BAC: $\overline{DE}/\overline{AC}=\overline{DB}/\overline{AB}=\overline{BE}/\overline{BC}$.
(2) △BXD and △BYA: $\overline{DB}/\overline{AB}=\overline{BX}/\overline{BY}=\overline{DX}/\overline{AY}$.
(3) △BXE and △BYC: $\overline{BX}/\overline{BY}=\overline{BE}/\overline{BC}=\overline{EX}/\overline{CY}$.

By substitution from (2) and (3) into (1)
we obtain, $\overline{DE}/\overline{AC}=\overline{DX}/\overline{AY}=\overline{EX}/\overline{CY}$.
By the SSS similarity property,
△DXE is similar to △AYC.

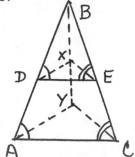

7. In figure 9.33, we are given a pyramid with triangular base and height H. Let there be a cross section parallel to the base h units from the top of the cone. Prove: Area △DXE/area △AYC=h^2/H^2.

From problem 6 we have △DXE similar to △AYC. Thus △DXE is a dilation image of △AYC. Let $D_{h/H,B}(△AYC)=△DXE$, where DE/AC = h/H. Therefore, Area △DXE/area△AYC= $(h/H)^2=h^2/H^2$.

8.   Volume of the larger pyramid, $V=(1/3)(9\sqrt{3})(12)=36\sqrt{3}$. $(B=s^2\sqrt{3}/4)$.
        Consider the smaller pyramid, P', a dilation image of the larger pyramid,
        P. By Theorem 9-4, $P'=r^3P$ where $r=4/12=1/3$ and B is the center of
        the dilation. Therefore, $V_{P'}=(1/3)^3V_P=(1/27)(36\sqrt{3})=4\sqrt{3}/3$.
        The frustrum of the pyramid has a volume, $V=36\sqrt{3} - 4\sqrt{3}/3=104\sqrt{3}/3$.

§ 4. Exercises
(Page 294)

1.   Given $(a+b)^3=a^3+3a^2b+3ab^2+b^3$,
        $(a+b)^4=a(a+b)^3+b(a+b)^3$
        $\quad\quad\quad=a(a^3+3a^2b+3ab^2+b^3)+b(a^3+3a^2b+3ab^2+b^3)$
        $\quad\quad\quad=a^4+3a^3b+3a^2b^2+ab^3+a^3b+3a^2b^2+3ab^3+b^4$
        $\quad\quad\quad=a^4+4a^3b+6a^2b^2+4ab^3+b^4$.

2. $(a+b)^5=a(a^4+4a^3b+6a^2b^2+4ab^3+b^4)+b(a^4+4a^3b+6a^2b^2+4ab^3+b^4)$
        $\quad\quad\quad=a^5+4a^4b+6a^3b^2+4a^2b^3+ab^4+ba^4+4a^3b^2+6a^2b^3+4ab^4+b^5$
        $\quad\quad\quad=a^5+5a^4b+10a^3b^2+10a^2b^3+5ab^4+b^5$.

CHAPTER 10

Vectors and Dot Product

§ 1. Exercises
(Pages 300-301)

1.   A+B=(1+3,4+2)=(4,6)

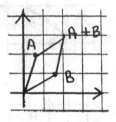

2.   A+B=(2,3)

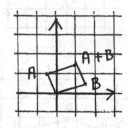

3.   A+B=(2,3)

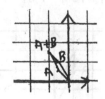

4.   A+B=(2,6)

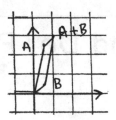

5.   A+B=(-1,3)

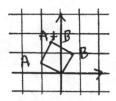

6.   A+B=(-4,-3)

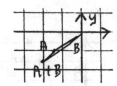

7.   (a)   A-B=(-2,2)

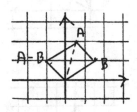

       (b)   A-B=(-4,1)

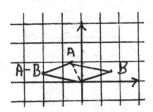

       (c)   A-B=(0,-1)

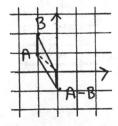

(d)   A-B=(0,4)

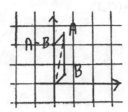

(e)   A-B=(-3,-1)

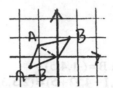

(f)   A-B=(-2,-1)

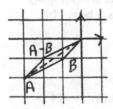

8.   Let A=(a_1,a_2) and B=(b_1,b_2).

(a)   cA=(ca_1,ca_2), cB=(cb_1,cb_2) and cA-cB=(ca_1-cb_1,ca_2-cb_2).

(b)   A-B=(a_1-b_1,a_2-b_2)

(c)   c(A-B)=c(a_1-b_1,a_2-b_2)
        =(c(a_1-b_1),c(a_2-b_2))
        =(ca_1-cb_1,ca_2-cb_2)
        =cA-cB by part (a).

9.   Let A=(1,2) and B=(3,1)
      A+B=(4,3)
      A+2B=(7,4)
      A+3B=(10,5)
      A-2B=(-5,0)
      A-3B=(-8,-1)
      A+4B=(13,6)
      A-4B=(-11,-2)

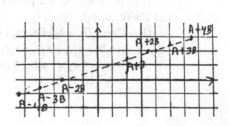

10.   Let $A=(2,-1)$ and $B=(-1,1)$
      $A+B=(1,0)$
      $A+2B=(0,1)$
      $A+3B=(-1,2)$
      $A+4B=(-2,3)$
      $A-B=(3,-2)$
      $A-2B=(4,-3)$
      $A-3B=(5,-4)$
      $A-4B=(6,-5)$
      $A+(1/2)B=(1.5,-0.5)$
      $A-(1/2)B=(2.5,-1.5)$

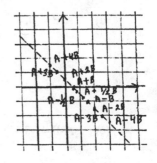

11.   (a) $A=(3,1,-2)$ and $B=(-1,4,5)$
            $A+B=(3+(-1),1+4,-2+5)=(2,5,3)$

      (b) $A=(a_1,a_2,a_3)$ and $B=(b_1,b_2,b_3)$
            $A+B=(a_1+b_1,a_2+b_2,a_3+b_3)$

§ 2. Exercise
(Page 304)

1.   (a) $A\cdot B=(1)(-1)+(3)(5)=14$

      (b) $A\cdot B=(-4)(-3)+(-2)(6)=0$

      (c) $A\cdot B=(3)(2)+(7)(-5)=-29$

      (d) $A\cdot B=(-5)(-4)+(2)(3)=26$

      (e) $A\cdot B=(-6)(-5)+(3)(-4)=18$

2.   Let $A=(a_1,a_2,a_3)$ and $B=(b_1,b_2,b_3)$
      $A\cdot B=a_1b_1+a_2b_2+a_3b_3$

      SP 1: It is commutative.
      $A\cdot B=a_1b_1+a_2b_2+a_3b_3 = b_1a_1+b_2a_2+b_3a_3 =B\cdot A$

      SP 2: If A, B and C are points, then distributivity holds.
      Let $C=(c_1,c_2,c_3)$.
      $A\cdot (B+C)=a_1(b_1+c_1)+a_2(b_2+c_2)+a_3(b_3+c_3)$
      $\qquad\qquad =a_1b_1+a_1c_1+a_2b_2+a_2c_2+a_3b_3+a_3c_3$
      $\qquad\qquad =(a_1b_1+a_2b_2+a_3b_3)+(a_1c_1+a_2c_2+a_3c_3)$
      $\qquad\qquad =A\cdot B + A\cdot C$

SP 3: If x is a number, then $(xA) \cdot B = x(A \cdot B) = A \cdot (xB)$.
$xA = (xa_1, xa_2, xa_3)$ and $xB = (xb_1, xb_2, xb_3)$
$$(xA) \cdot B = (xa_1)b_1 + (xa_2)b_2 + (xa_3)b_3$$
$$= a_1(xb_1) + a_2(xb_2) + a_3(xb_3) = A \cdot (xB)$$
$$= x(a_1b_1) + x(a_2b_2) + x(a_3b_3) = x(A \cdot B).$$

(a) $A \cdot B = (3)(-2) + (1)(3) + (-1)(4) = -7$

(b) $A \cdot B = (-4)(2) + (1)(1) + (1)(1) = -6$

(c) $A \cdot B = (3)(-2) + (-1)(4) + (5)(2) = 0$

(d) $A \cdot B = (4)(3) + (1)(2) + (-2)(7) = 0$

3. $(A+B)^2 = (A+B) \cdot (A+B)$
$$= A \cdot (A+B) + B \cdot (A+B) \qquad \text{by distributivity}$$
$$= A \cdot A + A \cdot B + B \cdot A + B \cdot B \quad \text{by distributivity}$$
$$= A^2 + A \cdot B + A \cdot B + B^2 \qquad \text{by commutativity}$$
$$= A^2 + 2A \cdot B + B^2$$

$(A+B) \cdot (A-B) = A \cdot (A-B) + B \cdot (A-B) \qquad \text{by distributivity}$
$$= A \cdot A - A \cdot B + B \cdot A - B \cdot B \quad \text{by distributivity}$$
$$= A^2 - A \cdot B + A \cdot B - B^2 \qquad \text{by commutativity}$$
$$= A^2 - B^2.$$

4.   (a) $|A| = [(5)(5) + (3)(3)]^{1/2} = \sqrt{34}$

(b) $|A| = \sqrt{34}$

(c) $|A| = \sqrt{34}$

(d) $|A| = \sqrt{34}$

(e) $|A| = \sqrt{20} = 2\sqrt{5}$

(f) $|A| = \sqrt{20} = 2\sqrt{5}$

(g) $|A| = \sqrt{20} = 2\sqrt{5}$

(h) $|A| = \sqrt{20} = 2\sqrt{5}$

5.   (a)   (b)

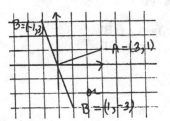   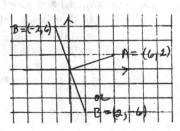

      (c)   (d)

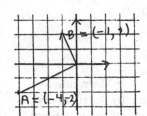   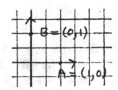

      (e)   (f)

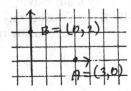   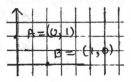

6.   (a)   2A=(6,2)

      (b)   2A=(12,4)

      (c)   2A=(-8,-4)

      (d)   2A=(2,0)

      (e)   2A=(6,0)

      (f)   2A=(0,2)

      OB is perpendicular to OP.

7. In each case, A · B=0 and P · B=0. Thus A · B=0 if and only if $\overline{OA}$ is perpendicular to $\overline{OB}$.

§ 3. Exercises

(Pages 308, 309)

1. $A \cdot B = (4)(-3) + (5)(1) = -7$, not perpendicular

2. $A \cdot B = (4)(-5) + (5)(4) = 0$, perpendicular

3. $A \cdot B = 0$, , perpendicular

4. $A \cdot B = 56$, not perpendicular

5. $A \cdot B = 0$, perpendicular

6. $A \cdot B = -56$, not perpendicular

7. $Q-P = (3-1, 6-5) = (2,1)$ and $N-M = (11-7, 4-2) = (4,2)$
    $(Q-P)(N-M) = (2)(4) + (1)(2) = 10$, They are not perpendicular.

8. $Q-P = (1,5)$ and $N-M = (-7,3)$ so $(Q-P)(N-M) = 8$ They are not perpendicular.

9. Given $P = (1,5)$, $Q = (3,6)$ and $M = (7,2)$, let $N = (n_1, n_2)$ and $N-M = (x_1, x_2)$.
    $|Q-P| = |(2,1)| = \sqrt{5}$.
    $|N-M| = |(x_1, x_2)| = \sqrt{x_1^2 + x_2^2}$ Thus,
    $x_1^2 + x_2^2 = 5$. Since PQ is perpendicular to MN,
    $(2,1) \cdot (x_1, x_2) = 2x_1 + x_2 = 0$ so $x_2 = -2x_1$.
    Solving these equations simultaneously, we have
    $$x_1^2 + x_2^2 = 5$$
    $$x_2 = -2x_1$$
    $$x_1^2 + (-2x_1)^2 = 5$$
    $$x_1^2 + 4x_1^2 = 5$$
    $$5x_1^2 = 5$$
    $$x_1^2 = 1 \text{ and } x_1 = 1 \text{ or } -1.$$
    If $x_1 = 1$, $x_2 = -2$. If $x_1 = -1$, $x_2 = 2$. Thus $N-M = (1,-2)$ or $(-1,2)$
    If $N-M = (1,-2)$, $N = (1,-2) + M = (1,-2) + (7,2) = (8,0)$.
    If $N-M = (-1,2)$, $N = (-1,2) + M = (-1,2) + (7,2) = (6,4)$.

10. Given $P = (-3,4)$, $Q = (-2,-1)$ and $M = (-4,-5)$, $Q-P = (1,-5)$.
    Let $N-M = (x_1, x_2)$.
    Since $(Q-P) \cdot (N-M) = 0$ and $|Q-P| = |N-M|$,
    we have $N-M = (5,1)$ or $(-5,-1)$.
    If $N-M = (5,1)$, $N = (5,1) + (-4,-5) = (1,-4)$.
    If $N-M = (-5,-1)$, $N = (-5,-1) + (-4,-5) = (-9,-6)$.

11.   Given P=(4,-1), Q=(5,3) and M=(7,-4), Q-P=(1,4).
       Let N-M=(x_1,x_2).
       Since (Q-P)·(N-M)=0 and $|Q-P|=|N-M|$,
       we have N-M=(4,-1) or (-4,1).
       If N-M=(4,-1), N=(4,-1)+(7,-4)=(11,-5).
       If N-M=(-4,1), N=(-4,1)+(7,-4)=(3,-3).

12.   Given P=(3,1), Q=(-1,1) and M=(2,-1), Q-P=(-4,0).
       Let N-M=(x_1,x_2).
       Since (Q-P)·(N-M)=0 and $|Q-P|=|N-M|$,
       we have N-M=(0,4) or (0,-4).
       If N-M=(0,4), N=(0,4)+(2,-1)=(2,3).
       If N-M=(0,-4), N=(0,-4)+(2,-1)=(2,-5)

13.   (a)   Given A ⊥ B and A ⊥ C,
              Prove: A ⊥ B+C.

              Since A ⊥ B, A·B=0. Also A ⊥ C implies A·C=0.
              By SP 2, A·(B+C)=A·B+A·C=0+0=0. Thus A ⊥ (B+C).

       (b)   Given A ⊥ B and C=xB for some number x,
              Prove: A ⊥ C.

              Since A ⊥ B, A·B=0.
              By SP 3, A·C=A·(xB)=x(A·B)=x(0)=0. Thus A ⊥ C.

14.   (a)   Let A=(2,3) and X=(-3,2) so that A·X=0.
              Choose B=(5,7) and C=B+X=(2,9). Thus,
              A·B=(2)(5)+(3)(7)=31.
              A·C=(2)(2)+(3)(9)=31; but B is not equal to C.

       (b)   If A·X=0, A·(B+X)=A·B+A·X=A·B+0=A·B. This idea is used
              in part (a).

15.   Given A ⊥ B, Prove: $|A+B|^2=|A|^2+|B|^2$.

       Since A⊥B implies A·B=0,
       $|A+B|^2=(A+B)^2=A^2+2A·B+B^2=A^2+B^2=|A|^2+|B|^2$.

16.   (a)   Given any A and B, prove: $|A+B|^2+|A-B|^2=2|A|^2+2|B|^2$.

              $|A+B|^2+|A-B|^2=(A^2+2A·B+B^2)+(A^2-2A·B+B^2)$
              $=2A^2+2B^2=2|A|^2+2|B|^2$.

(b)

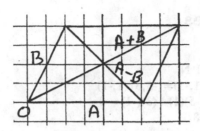

The sum of the squares of the diagonals equals twice the sum of the squares of the sides.

17.   Given any A and B, prove $|A+B|^2-|A-B|^2=4A\cdot B$.

$|A+B|^2-|A-B|^2=(A^2+2A\cdot B+B^2)-(A^2-2A\cdot B+B^2)=4A\cdot B$.

18.   Given $A \perp B$ and neither A nor B equal 0, let x, y be numbers such that $xA+yB=0$. Prove: $x=y=0$.

Since $xA+yB=0$, $A\cdot(xA+yB)=A\cdot 0=0$ so $xA^2+yA\cdot B=0$.
Given $A \perp B$, $A\cdot B=0$. By SP4, if A is non-zero, $A^2>0$.
Thus, $xA^2+yA\cdot B=xA^2+0=xA^2=0$ and $x=0$.
Similarly, $(xA+yB)\cdot B=xA\cdot B+yB^2=0\cdot B=0$. Again $A\cdot B=0$ and $B^2>0$ so $xA\cdot B+yB^2=0+yB^2=yB^2=0$ and $y=0$.

§ 4. Exercises
(Page 311)

1.   (a) $B=(5,7)$, $A=(3,-4)$ $c=A\cdot B/B\cdot B=[(3)(5)+(-4)(7)]/[5^2+7^2]=-13/74$

      (b) $B=(2,-1)$, $A=(-3,-4)$ $c=[(-3)(2)+(-4)(-1)]/[2^2+(-1)^2]=-2/5$

      (c) $B=(-3,-2)$, $A=(5,1)$ $c=[(5)(-3)+(1)(-2)]/[(-3)^2+(-2)^2]=-17/13$

      (d) $B=(-4,1)$, $A=(2,-5)$ $c=[(2)(-4)+(-5)(1)]/[(-4)^2+1^2]=-13/17$

2.   (a) $cB=(-13/74)(5,7)=(-65/74,-91/74)$

      (b) $cB=(-2/5)(2,-1)=(-4/5,2/5)$

      (c) $cB=(-17/13)(-3,-2)=(51/13,34/13)$

      (d) $cB=(-13/17)(-4,1)=(52/17,-13/17)$

§ 5. Exercises
(Pages 315,316)

1.   (a) $X \cdot N = P \cdot N$
              $(x,y) \cdot (1,-1) = (-5,3) \cdot (1,-1)$
              $x - y = (-5)(1) + (3)(-1) = -8$
              $x - y = -8$

   (b) $(x,y) \cdot (-3,5) = (7,-2) \cdot (-3,5)$
              $-3x + 5y = -31$

2.   (a) $(x,y) \cdot (-5,4) = (3,2) \cdot (-5,4)$
              $-5x + 4y = -7$

   (b) $(x,y) \cdot (2,8) = (1,1) \cdot (2,8)$
              $2x + 8y = 10$

3.   (a) $(x,y) \cdot (4,1) = (5,-2) \cdot (4,1)$
              $4x + y = 18$

   (b) $(x,y) \cdot (-5,3) = (6,-7) \cdot (-5,3)$
              $-5x + 3y = -51$

4. $L_1: 3x - 5y = 1$ so $N_1 = (3,-5)$
      $L_2: 2x + 3y = 5$ so $N^2 = (2,3)$
      $N_1 \cdot N_2 = (3,-5) \cdot (2,3) = (3)(2) + (-5)(3) = -9$
      Thus N_1 is not perpendicular to N_2 so L_1 is not perpendicular to L_2.

5.   (a) $(3,-5) \cdot (2,1) = 1 \neq 0$ so L_1 is not perpendicular to L_2.

   (b) $(2,7) \cdot (1,-1) = -5 \neq 0$ so L_1 is not perpendicular to L_2.

   (c) $(3,-5) \cdot (5,3) = 0$ Thus $L_1 \perp L_2$.

   (d) $(-1,1) \cdot (1,1) = 0$ Thus, $L_1 \perp L_2$.

6.   (a) $m_1 = -1$ and $m_2 = 3$, $m_1 \cdot m_2 = (-1)(3) = -3$. Thus not perpendicular.

   (b) $m_1 = 3$ and $m_2 = -1/3$, $m_1 \cdot m_2 = (3)(-1/3) = -1$. Thus $L_1 \perp L_2$.

  (c) $m_1 = 4$ and $m_2 = -1/4$, $m_1 \cdot m_2 = (4)(-1/4) = -1$. Thus $L_1 \perp L_2$.

   (d) $m_1 = 1/2$ and $m_2 = 2$, $m_1 \cdot m_2 = (1/2)(2) = 1$. Thus not perpendicular.

§ 6. Exercises
(Pages 318,319)

1.　(a)　$A=(3,-1,4)$ and $B=(1,2,-1)$
$c=A\cdot B/B\cdot B=(3,-1,4)\cdot(1,2,-1)/(1,2,-1)\cdot(1,2,-1)$
$c=[(3)(1)+(-1)(2)+(4)(-1)]/[1^2+2^2+(-1)^2]=-3/6=-1/2.$

　　(b)　$c=[(1)(-1)+(2)(-2)+(-1)(-5)]/[(-1)^2+(-2)^2+(-5)^2]=0/30=0$

　　(c)　$c=[(3)(2)+(-1)(-2)+(2)(1)]/[2^2+(-2)^2+1^2]=10/9$

　　(d)　$c=[(-4)(-1)+(3)(3)+(1)(-2)]/[(-1)^2+3^2+(-2)^2]=11/14$

　　(e)　$c=[(1)(-4)+(1)(2)+(1)(1)]/[(-4)^2+2^2+1^2]=-1/21$

2.　(a)　$cB=(-1/2)(1,2,-1)=(-1/2,-1,1/2)$

　　(b)　$cB=(0)(-1,-2,-5)=(0,0,0)$

　　(c)　$cB=(10/9)(2,-2,1)=(20/9,-20/9,10/9)$

　　(d)　$cB=(11/14)(-1,3,-2)=(-11/14,33/14,-22/14)$

　　(e)　$cB=(-1/21)(-4,2,1)=(4/21,-2/21,-1/21)$

3. Check your proofs and see that each step also holds in 3-dimensions.

4.　(a)　$|A|^2=3^2+(-1)^2+4^2=26$ so $|A|=\sqrt{26}.$

　　(b)　$|A|^2=1^2+2^2+(-1)^2=6$ so $|A|=\sqrt{6}.$

　　(c)　$|A|^2=3^2+(-1)^2+2^2=12$ so $|A|=\sqrt{14}.$

　　(d)　$|A|^2=(-4)^2+3^2+1^2=26$ so $|A|=\sqrt{26}.$

　　(e)　$|A|^2=1^2+1^2+1^2=3$ so $|A|=\sqrt{3}.$

5.　(a)　$A=(3,-1,4)$, $B=(1,2,-1)$ and $A-B=(2,-3,5)$
$d(A,B)=|A-B|$ so $|A-B|^2=2^2+(-3)^2+5^2=38$ so $d(A,B)=\sqrt{38}.$

(b) $A=(1,2,-1)$, $B=(-1,-2,-5)$ and $A-B=(2,4,4)$
$|A-B|^2=2^2+4^2+4^2=36$ so $d(A,B)=6$.

(c) $A=(3,-1,2)$, $B=(2,-2,1)$ and $A-B=(1,1,1)$
$|A-B|^2=1^2+1^2+1^2=3$ so $d(A,B)=\sqrt{3}$.

(d) $A=(-4,3,1)$, $B=(-1,3,-2)$ and $A-B=(-3,0,3)$
$|A-B|^2=(-3)^2+0^2+3^2=18$ so $d(A,B)=3\sqrt{2}$.

(e) $A=(1,1,1)$, $B=(-4,2,1)$ and $A-B=(5,-1,0)$
$|A-B|^2=5^2+(-1)^2+0^2=26$ so $d(A,B)=\sqrt{26}$.

6.   Let A, B, and C be non-zero elements of R^3 which are mutually perpendicular. Let x, y, and z be numbers such that $xA+yB+zC=0$. Prove: $x=y=z=0$.

Dot both sides of $xA+yB+zC=0$ by A obtaining
$A{\cdot}(xA+yB+zC)=A{\cdot}0$
$x(A{\cdot}A)+y(A{\cdot}B)+z(A{\cdot}C)=0$.   Since $A\perp B$ and $A\perp C$, $A{\cdot}B=A{\cdot}C=0$.
Thus, $xA^2=0$. Since A is non-zero, $|A|>0$. Therefore $x=0$.
If one dots both sides by B, it follows that $y=0$.
If one dots both sides by C, it follows that $z=0$.
Therefore, $x=y=z=0$.

7.   Let $A=(a_1,a_2,a_3)$, $E_1=(1,0,0)$, $E_2=(0,1,0)$ and $E_3=(0,0,1)$.
$A{\cdot}E_1=a_1(1)+a_2(0)+a_3(0)=a_1$.
$A{\cdot}E_2=a_1(0)+a_2(1)+a_3(0)=a_2$.
$A{\cdot}E_3=a_1(0)+a_2(0)+a_3(1)=a_3$.

8.   Let $A=(a_1,a_2,a_3,a_4)$ and $B=(b_1,b_2,b_3,b_4)$ be points in R^4.
$A+B=(a_1+b_1,a_2+b_2,a_3+b_3,a_4+b_4)$
Let x be a number, $xA=(xa_1,xa_2,xa_3,xa_4)$
$A{\cdot}B=a_1b_1+a_2b_2+a_3b_3+a_4b_4$
For SP1 it is clear that each pair commutes as real numbers so SP 1 holds for the dot product in R^4 (and in R^n).
SP2 follows since the real number distributive property holds in the products in the dot product definition for all R^n.
SP3 follows since the real number products are associative and commutative. This holds in all R^n.
SP4 follows from the observation that the squares of non-zero real numbers are greater than 0. Thus, if a non-zero A is in R^n, $A{\cdot}A>0$. If $A=0$, all of the squares are zero so $A{\cdot}A=0$.
Therefore, SP1-SP4 hold in all R^n.

§ 7. Exercises
(Page 320)

1. (a) $X \cdot N = P \cdot N$ where $P = (-4,5,1)$ and $N = (2,3,1)$
$(x,y,z) \cdot (2,3,1) = (-4,5,1) \cdot (2,3,1) = -8 + 15 + 1 = 8$
$2x + 3y + z = 8$

(b) $(x,y,z) \cdot (-2,1,4) = (1,-3,-7) \cdot (-2,1,4)$
$-2x + y + 4z = -33$

(c) $(x,y,z) \cdot (1,-1,1) = (-2,1,3) \cdot (1,-1,1)$
$x - y + z = 0$

(d) $(x,y,z) \cdot (2,1,3) = (3,1,-1) \cdot (2,1,3)$
$2x + y + 3z = 4$

2. (a)   Let P_1: $3x - 2y + 5z = 0$ and P_2: $4x - y + 7z = 1$
Let $N_1 = (3,-2,5)$ and $N_2 = (4,-1,7)$
$N_1 \cdot N_2 = (3)(4) + (-2)(-1) + (5)(7) = 49$.
Since $N_1 \cdot N_2 > 0$, N_1 is not perpendicular to N_2 and P_1 is not perpendicular to P_2.

(b) $N_1 \cdot N_2 = (3,-2,5) \cdot (7,-3,2) = 37$. Thus P_1 is not perpendicular to P_2.

(c) $N_1 \cdot N_2 = (3,-2,-5) \cdot (6,4,2) = 0$. Thus $P_1 \perp P_2$.

(d) $N_1 \cdot N_2 = (3,4,11) \cdot (2,-7,2) = 0$. Thus $P_1 \perp P_2$.
(e) $N_1 \cdot N_2 = (7,-1,8) \cdot (8,2,-1) = 46$. Thus P_1 is not perpendicular to P_2.

CHAPTER 11

Transformations

§ 1. Exercises
(page 323)

1. (a) $|XZ|^2 = 10^2 + 24^2 = 676$ so $|XZ| = 26$

   (b) The sides of $\triangle XYZ$ are twice the lengths of the corresponding sides of $\triangle ABC$.

   (c) $P(\triangle ABC) = 5 + 12 + 13 = 30$ cm and $P(\triangle XYZ) = 10 + 24 + 26 = 60$ cm. The perimeter of $\triangle XYZ$ is twice the perimeter of $\triangle ABC$.

   (d) $A(\triangle ABC) = (1/2)(5)(12) = 30$ cm^2 and $A(\triangle XYZ) = (1/2)(10)(24) = 120$ cm^2. The area of $\triangle XYZ$ is four times the area of $\triangle ABC$.

   (e) The corresponding angles of the two triangles have the same measure.

2. (a) TOT, HAH and TOOT. (There are other correct choices.)

   (b) These words work because each letter has a vertical axis of symmetry and the word reads the same backwards and forwards.

Experiment 11-2
(Pages 326,327)

1.

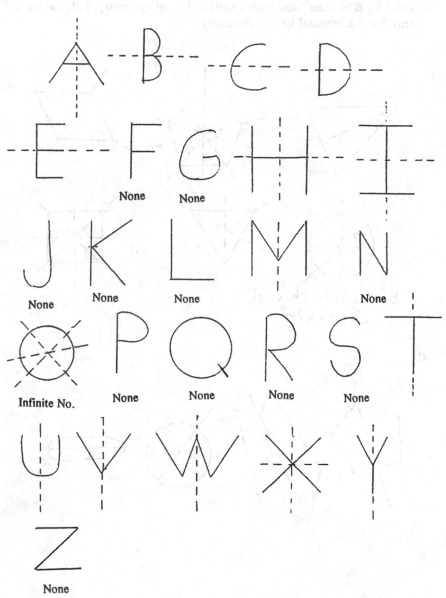

2. You cannot read **TOO** in a mirror. You cannot read **EYE** in the mirror. The word **MOM** has letters with a vertical line of symmetry and the word has a vertical line of symmetry. The word **TOO** has letters with a vertical line of symmetry; but, the word does not have a vertical line of symmetry. The word **EYE** has one letter with a vertical line of symmetry; but, the letter E only has a horizontal line of symmetry.

3.

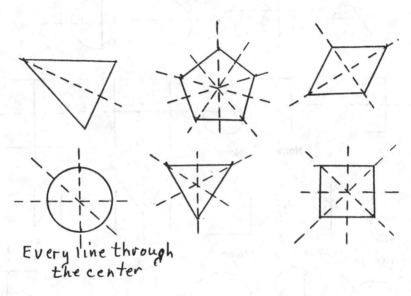

Every line through
the center

4.

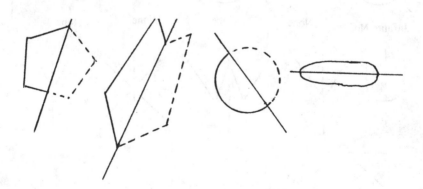

5. A figure has line symmetry if each point can be paired with another point such that the segment joining these points is perpendicular to the line of symmetry and the midpoint of the segment lies on the line of symmetry.

§ 3. Exercises
(Pages 330, 331)

1. TOOT TOT TOOT, WOW MOM WOW, WOW TOT WOW, ...

2. Write up your findings.

3. Read this chapter in <u>One, Two, Three ... Infinity</u>.

4. (a)

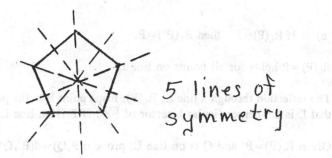

5 lines of symmetry

(b)

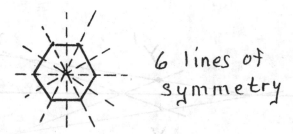

6 lines of symmetry

(c)

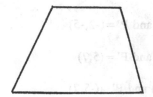

§ 4. Exercises

(Pages 334-337)

1.   (a)

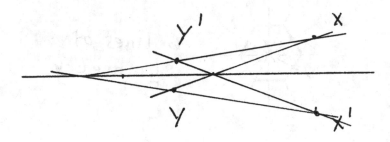

(b)   If $R_L(P)=P'$, then $R_L(P')=P$.

2. $R_L(P)=P$ holds for all points on line L.

3.   The reflection through a line L, $R_L(P)$, maps point P to the point P' such that L is the perpendicular bisector of $\overline{PP'}$. If P is on line L, $R_L(P)=P$.

4.   Given $R_L(P)=P'$ and Q is on line L, prove $d(P,Q)=d(P',Q)$.

Since $R_L(P)=P'$, by the definition in exercise 3 we know L is the perpendicular bisector of $\overline{PP'}$. Therefore, by the Perpendicular Bisector Theorem (Theorem 5-1) we conclude that $d(P,Q)=d(P',Q)$.

5.

6.   Using the x-axis as the line of reflection:

(a) $P=(3,7)$ and $P'=(3,-7)$

(b) $P=(-2,5)$ and $P'=(-2,-5)$

(c) $P=(5,-7)$ and $P'=(5,7)$

(d) $P=(-5,-7)$ and $P'=(-5,7)$

    (e) $P=(p_1,p_2)$ and $P'=(p_1,-p_2)$

7.   Using the y-axis as the line of reflection:

    (a) $P=(3,7)$ and $P'=(-3,7)$

    (b) $P=(-2,5)$ and $P'=(2,5)$

    (c) $P=(5,-7)$ and $P'=(-5,-7)$

    (d) $P=(-5,-7)$ and $P'=(5,-7)$

    (e) $P=(p_1,p_2)$ and $P'=(-p_1,p_2)$

8.   When the line $x=3$ is reflected in the x-axis it is its own image. In other words, if p is on the line $x=3$, $R_x(P)=P'$ where P' is also on the line $x=3$.

9.   The image of the line $x=3$ when reflected in the $y=$axis is the line $x=-3$.

10.   In figure 11.25, let L be the line corresponding to the river bank. Let $R_L(A)=A'$ and let X be the intersection of $\overline{A'B}$ with L. From the SEG postulate we obtain:
$|A'X|+|XB|=|A'B|$. From exercise 4 we have $|A'X|=|AX|$. By substitution we have $|AX|+|XB|=|A'B|$.
Let Y be any other point on L. Since A', B and Y are non-collinear, the Triangle Inequality Postulate implies $|A'Y|+|YB|>|AB|$. By exercise 4 we obtain $|AY|=|A'Y|$. By substitution, $|AY|+|YB|>|A'B|$. Therefore, any location different from X will cause the driver to drive further than the path A to X to B.

11.   In figure 11.26 R_L is the reflection in line L. Also, $R_L(P)=P'$ and $R_L(Q)=Q'$. Prove: $d(P,Q)=d(P',Q')$.

Draw the line through P perpendicular to $\overline{QQ'}$ at R and another line through P' perpendicular to $\overline{QQ'}$ at R'. Thus, $\overline{PR}$ is parallel to $\overline{P'R'}$. Since $\overline{PP'}$ and $\overline{QQ'}$ are perpendicular to line L, they are parallel. A parallelogram with one right angle is a rectangle. Thus PRR'P' is a rectangle and $|PR|=|P'R'|$. Since $R_L(P)=P'$, L is the perpendicular bisector of $\overline{PP'}$ and is also the perpendicular bisector of $\overline{RR'}$. In addition, L is the perpendicular bisector of $\overline{QQ'}$. Let Y be the intersection of L with the line $\overline{QQ'}$. From each of the above we obtain, $|RY|=|R'Y|$ and $|QY|=|Q'Y|$. By the SEG Postulate
$|QY|=|QR|+|RY|$ and
$|Q'Y|=|Q'R'|+|R'Y|$. By substitution,

$|QR|+|RY|=|Q'R'|+|R'Y|$. Since $|RY|=|R'Y|$ we can use the subtraction property of equality and obtain $|QR|=|Q'R'|$. In the right triangles $\triangle PQR$ and $\triangle P'Q'R'$ we have $|PR|=|P'R'|$ and $|QR|=|Q'R'|$. By the Right Triangle Postulate $|PQ|=P'Q'|$.

12.   Let $P=(p_1,p_2,p_3)$ be a point in three-space.

   (a) $R_{xy}(P)=(p_1,p_2,-p_3)$

   (b) $R_{xz}(P)=(p_1,-p_2,p_3)$

   (c) $R_{yz}(P)=(-p_1,p_2,p_3)$

§ 5. Exercises
(Pages 339-340)

1.

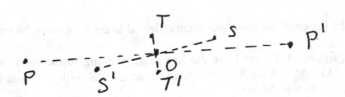

2.   (a)

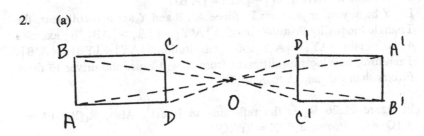

   (b)

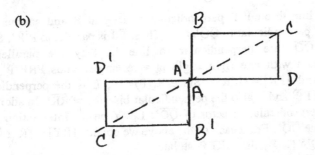

   (c) $R_X(ABCD)=CDAB$

3. A reflection through a point O, $R_O(P)$, pairs P with a point -P such that O is the midpoint of the segment P(-P).

4. If the origin is the point of the reflection, O, then -A is:

    (a)   -A=(7,-56)

    (b)   -A=(4,7)

    (c)   -A=(-3,8)

5. When the line x=4 is reflected in the origin, the image is x=-4.

6. Let $P=(p_1,p_2,p_3)$ in 3-space and the origin be the point of reflection, then $R_O(P)=(-p_1,-p_2,-p_3)$.

§ 6. Exercises
(Pages 344-346)

1.

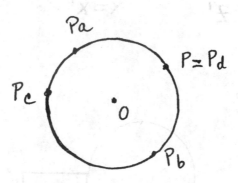

2.   (a) $G_{400}=G_{40}$

    (b) $G_{-75}=G_{285}$

    (c) $G_{1080}=G_0$

    (d) $G_{-500}=G_{220}$

    (e) $G_{-780}=G_{300}$

3.   (a) $G_{400}(P)=P_a$

      (b) $G_{-75}(P)=P_b$

      (c) $G_{1080}(P)=P_c$

      (d) $G_{-500}(P)=P_d$

      (e) $G_{-780}(P)=P_e$

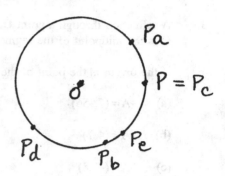

4. $G_{120,X}(\triangle ABC)=\triangle A'B'C'$

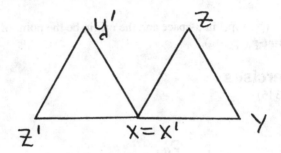

5. $G_{270}=G_{-90}$

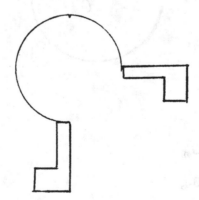

6.   Let X be a point in the plane. $G_{180}(X)=X'$ where $\triangle XOX'$ is a straight angle. O is the midpoint of the segment XX' so $R_O(X)=X'$. Since $G_{-180}=G_{180}$ it follows that $G_{-180}(X)=X'=R_O(X)$.

7. $I = G_{360k}$ where k is an integer.

8.   Let x and y be numbers and $G_{x,0}=G_{y,0}$. Then x=y+360k where k is an integer.

9.   Let x be a number greater than 360. To find y ($360 \geq y \geq 0$), find the largest multiple, n, such that 360n<x. Then y=x-360n.

10.   Let P=(4,0).

      (a) $G_{90}(P)=(0,4)$

      (b) $G_{180}(P)=(-4,0)$

      (c) $G_{-270}(P)=(0,4)$

      (d) $G_{360}(P)=(4,0)$

11.   Let P=(0,-6)

      (a) $G_{90}(P)=(6,0)$

      (b) $G_{180}(P)=(0,6)$

      (c) $G_{-270}(P)=(6,0)$

      (d) $G_{360}(P)=(0,-6)$

12.   Let P=(3,6)

      (a) $G_{90}(P)=(-6,3)$

      (b) $G_{180}(P)=(-3,-6)$

      (c) $G_{-270}(P)=(-6,3)$

      (d) $G_{360}(P)=(3,6)$

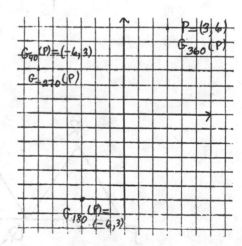

13.   Let $\triangle$PQR have vertices P=(3,2), Q=(3,-2) and R=(6,0).

      (a)   When reflected through the y-axis we obtain P'=(-3,2), Q'=(-3,-2) and R'=(-6,0)

      (b)   The rotation $G_{180,0}(\triangle PQR)=\triangle Q'P'R'$. $G_{-180,0}$ would also work.

14. (a) The regular hexagon has 6 lines of symmetry.

   (b) $G_{120,0}(P_1)=P_3$.

   (c) $G_{720/n,0}(P_1)=P_3$. Adjacent points in an n-gon form an angle with O at the vertex equal to 360°/n. The rotation from P_1 to P_3 is equal to two of these angles, thus 720/n.

   (d) $G_{-120,P2}(P_1)=P_3=G_{240,P2}(P_1)$.

§ 7. Exercises
(Pages 347, 348)

1.

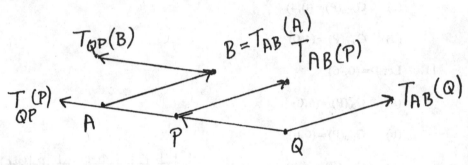

2.

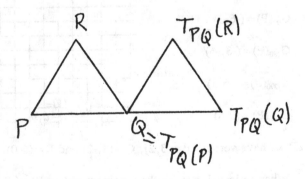

3.

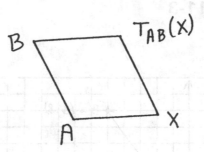

A parallelogram

4.

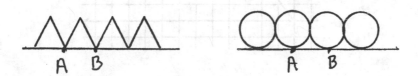

5. Let T be a translation.
    Prove: $d(P,Q)=d(T(P),T(Q))$

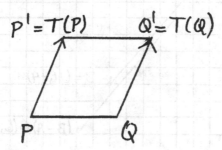

Using the definition of translation, PP'||QQ' and $d(P,P')=d(Q,Q')$. By
Theorem 7-5, since we have one pair of sides of the quadrilateral
PQTP'Q' both parallel and of the same length, we have PQTP'Q' is a
parallelogram. Using Theorem 7-1, the opposite sides of a parallelogram
are of the same length. Therefore, $d(P,Q)=d(P',Q')=d(T(P),T(Q))$.

Experiment 11-3
(Pages 348,349)

1.

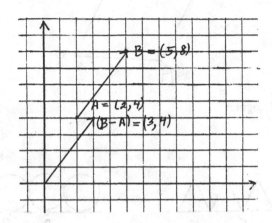

2.

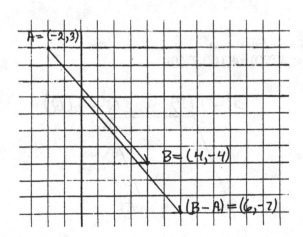

3.   The points O,A,B and (B-A) are the vertices of a parallelogram.

4. $d(A,B)=d(O,(B-A))-$

5.   AB and O(B-A) have the same direction.

6. $T_{AB}=T_{O(B-A)}$

§ 8. Exercises
(Pages 352,353)

1. $A=(-2,5)$ Let $P'=T_A(P)=P+A$.

   (a) If $P=(1,5)$, then $P'=(1,5)+(-2,5)=(-1,10)$

   (b) If $P=(-3,-6)$, then $P'=(-5,-1)$.

   (c) If $P=(2,-5)$, then $P'=(0,0)$.

   (d) If $P=(0,0)$, then $P'=(-2,5)$.

   (e) If $P=(p_1,p_2)$, then $P'=(-2+p_1,5+p_2)$.

2.   Let $A=(-3,2)$. Let S be a triangle with vertices $(2,5)$, $(-3,7)$, and $(3,6)$.
   $T_A(2,5)=(2,5)+(-3,2)=(-1,7)$
   $T_A(-3,7)=(-6,9)$ and
   $T_A(3,6)=(0,8)$.
   Thus, $T_A(S)$ is a triangle with vertices $(-1,7)$, $(-6,9)$, and $(0,8)$.

3.   Let $A=(-3,2)$ and L be the line $x=4$.
   $T_A(L)$ is the line $x=1$.

4.   Let $A=(-3,2)$ and K be the line $y=4$.
   $T_A(K)$ is the line $y=6$.

5.   Let $A=(-3,2)$ and C be a circle centered at the origin with radius 3.
   $T_A(C)$ is a circle centered at $(-3,2)$ with radius 3.

6.   Let $A=(0,0)$, $B=(2,3)$, $C=(4,6)$, $D=(-2,-3)$, $E=(3,2)$, and $F=(5,5)$.

   (a) $T_{AB}=T_{BA}$ is false since $B-A=(2,3)$ and $A-B=(-2,-3)$.

   (b) $T_{AB}=T_{BC}$ is true since $B-A=(2,3)$ and $C-B=(2,3)$.

   (c) $T_{BC}=T_{AD}$ is false since $C-B=(2,3)$ and $D-A=(-2,-3)$.

   (d) $T_{BC}=T_{DA}$ is true since $C-B=(2,3)$ and $A-D=(2,3)$.

   (e) $T_{AB}=T_{EC}$ is false since $B-A=(2,3)$ and $C-E=(1,4)$.

   (f) $T_{BC}=T_{EF}$ is true since $C-B=(2,3)$ and $F-E=(2,3)$.

(g) $T_{EC} = T_{EF}$ is false since C-E=(1,4) and F-E=(2,3).

(h) $T_{DB} = T_{AC}$ is true since B-D=(4,6) and C-A=(4,6).

Additional Exercises for Chapter 11
(Pages 353-355)

1.   (a)   For the identity mapping, all points of the plane are fixed points.

       (b) R_O has the point O as its only fixed point.

       (c) R_L has the line L as the set of fixed points.

       (d) $G_{x,0}(P)$ has one fixed point, point O.

       (e)   A translation has no fixed points unless it is of 0 length in which case the entire plane of points is fixed.

       (f)   The constant mapping whose value is a given point X has X as its only fixed point.

2.

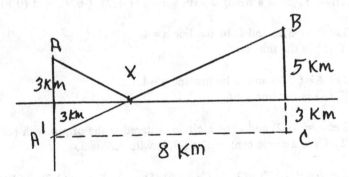

Note right triangle $\triangle A'BC$ where $|A'B|^2 = 8^2 + 8^2 = 128$ so $|A'B| = 8\sqrt{2}$. Observing that $|AX| = |A'X|$ and $|A'B| = |A'X| + |XB|$ we obtain by substitution that $|AX| + |XB| = 8\sqrt{2}$ km.

3.   Use your creativity to make up three mappings.

4.   (a) $R_{x\text{-axis}}(3,3) = (-3,3) = A'$

5. $T_{AB}(6,3) = (9,3) = B'$

6. $G_{90,B}(9,3)=(6,6)=B''$

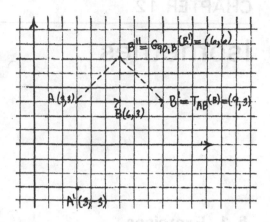

7. $d(A',B'')^2=d[(3,-3),(6,6)]^2=(6-3)^2+(6-(-3))^2=9+81=90$
      $d(A',B'')=3\sqrt{10}$.

8.   Area of $\triangle AB''B' = (1/2)(6)(3)=9$ sq. units.

9. $m(\angle AB''B')=90°$.

10.   In figure 11.35, reflect A through line L_1 and reflect B through line L_2. Let X be the point where $\overline{A'B}$ intersects L_1 and Y be the point where $\overline{AB'}$ intersects L_2. The path would be A to X to Y to B.

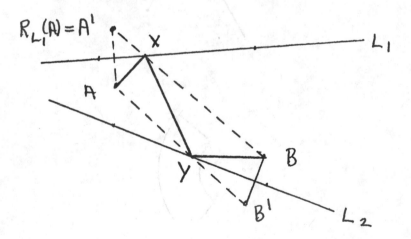

CHAPTER 12

Isometries

§ 1. Exercises
(Pages 360-363)

1.

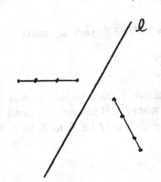

2.

3.

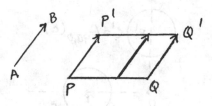

4.   (a)

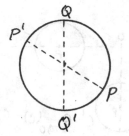

Note that $R_P(C) = C$ but each point maps to the point on the other side.

(b)

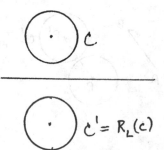

(c)

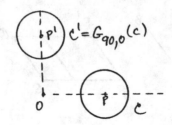

$c' = G_{90,0}(c)$

(d)

$c' = G_{270,0}(c) = G_{-90,0}(c)$

(e)

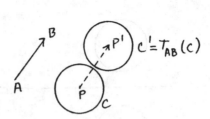

$c' = T_{AB}(c)$

5.   (a)

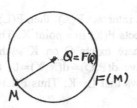

(b)   Since Q is a fixed point and the isometry preserves the distance
       d(M,Q) the possible locations for M lie on the circle centered at Q
       with radius d(M,Q).

6.   (a)   _____
             F(P)=P   F(M)=M   F(Q)=Q

      (b)   Since F is an isometry,
             d(P,M)=d(F(P),F(M)),
             d(M,Q)=d(F(M),F(Q)), and
             d(P,Q)=d(F(P),F(Q)). Since M is on PQ, we have
             d(P,M)+d(M,Q)=d(P,Q) by the SEG postulate. By substitution,
             d(F(P),F(M))+d(F(M),F(Q))=d(F(P),F(Q)).
             Therefore, F(M) must lie on PQ at M.

7.   (a)   _____
             F(P)=P   F(Q)=Q   F(M)=M

      (b)   Since F is an isometry,
             d(P,M)=d(F(P),F(M)),
             d(M,Q)=d(F(M),F(Q)), and
             d(P,Q)=d(F(P),F(Q)). Since Q is on PM, we have
             d(P,Q)+d(Q,M)=d(P,M) by the SEG postulate. By substitution,
             d(F(P),F(Q))+d(F(Q),F(M))=d(F(P),F(M)).
             Therefore, F(M) is on the segment PM at M.

8.   Given that line L is parallel to line K and that F is an isometry,
      Prove: F(L)||F(K).

      If F(L) does not intersect F(K), then F(L) and F(K) are parallel. Suppose
      that F(L) intersects F(K) in a point X. Thus there exists P on L such that
      X=F(P) and there exists Q on K such that X=F(Q). Since F is an
      isometry we have d(P,Q)=d(X,X)=0, thus P=Q. Since L and K are
      parallel it follows that L=K. Thus F(L)=F(K) and F(L)||F(K).

9.   Let L and K be perpendicular lines, and let F be an isometry.
      Prove: F(K) and F(L) are perpendicular lines.

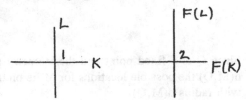

      By ISOM 2, the image of a line is a line. Thus line L goes to line F(L)
      and line K goes to line F(K) and their common point X goes to the
      common point of F(K) and F(L), point F(X).
      Since K is perpendicular to L, m($\angle$1)=90°. The image of angle 1 is
      angle 2 thus m($\angle$2)=90° because isometries preserve angle measure.
      Therefore, F(L) is perpendicular to F(K).

10.   Let R be a rectangle with vertices A, B, C, and D. Let F be an isometry.
      Prove: The area of R equals the area of F(R).

      By problem 9 perpendicular lines are mapped by an isometry to
      perpendicular lines. Thus the right angles formed by adjacent sides in R
      are mapped to right angles formed by corresponding adjacent sides in
      F(R). Consequently, F(R) is a rectangle. The area of R equals
      |AD||DC|. Since F is an isometry, d(A,D)=d(F(A),F(D)) and
      d(D,C)=d(F(D),F(C)). By substitution we obtain that area of R equals
      the area of F(R).

11.   In figure, 12.7, let PQ and MN meet in a point) and bisect each other.
      Prove: d(P,M)=d(Q,N).

      Let R_0 be a reflection through O. Thus R_0(P)=Q and angle $\angle$POM maps
      to $\angle$QON. Since $\angle$OPM maps to $\angle$OQN we know that the intersection
      of lines $\overleftrightarrow{PM}$ and $\overleftrightarrow{OM}$, M, maps to the intersection of the lines $\overleftrightarrow{QN}$ and
      $\overleftrightarrow{ON}$, N. Therefore, d(P,M)=d(Q,N).

12. In figure 12.8, let $|WO|=|YO|$, $|VO|=|VW|$, and $m(\angle VOW)=m(\angle ZOY)$. Prove: $|ZY|=|VW|$.

In figure 12.8, let $m(\angle ZOY)=n°$ and $m(\angle YOV)=p°$. Let $G_{(n+p),0}$ be the rotation around O of $(n+p)°$. Thus, $G(OZ)=OV$ and $G(OY)=OW$ since angles are preserved and $m(\angle VOW)=m(\angle ZOY)$. Since length is preserved $G(Z)=V$ and $G(Y)=W$ and $|ZY|=|VW|$.

13. Given that line L is the perpendicular bisector of $\overline{PQ}$ and $\overline{XY}$, Prove: $d(P,X)=d(Q,Y)$.

Let R be a reflection through the line L. By definition, $R(P)=Q$ and $R(X)=Y$. A reflection is an isometry so $d(P,X)=d(R(P),R(X))$. By substitution, $d(P,X)=d(Q,Y)$.

14. (a) If $A=(-8,9)$ and $B=(1,3)$, $P_y(A)=(0,9)$ and $P_y(B)=(0,3)$.

    (b) P_y is not an isometry since $d(A,B)$ is not equal to $d(R_y(A),R_y(B))$.

    (c) If $X=(x,y)$ and $V=(u,v)$ and $P_y(X)=X'$ and $p_y(V)=V'$, then $d(X',V')=\sqrt{(y-v)^2}=|y-v|$.

    (d) If A and B are points in the plane and $P_y(A)=P_y(B)$, A and B have the same y-coordinate,b, thus they lie on the line $y=b$.

    (e) If T is a triangle in the plane, the image of T under P_y is a line segment in the y-axis.

§ 2. Exercises
(Page 366)

1. (a) $P=(-1,3)$ so $|P|^2=(-1)^2+3^2=10$ and $|P|=\sqrt{10}$.

    (b) $P=(137,137)$ so $|P|^2=137^2+137^2=2(137)^2$ and $|P|=137\sqrt{2}$.

    (c) $P=(1,-3)$ so $|P|=\sqrt{10}$.

2. (a) $A=(3,6)$ and $B=(3,-4)$ so $|A-B|^2=(3-3)^2+(6-(-4))^2=100$. $|A-B|=10$.

    (b) $A=(-2,4)$ and $B=(-3,-5)$ so $|A-B|^2=(-2-(-3))^2+(4-(-5))^2=82$. $|A-B|=\sqrt{82}$.

(c) $A=(0.5,7)$ and $B=(-1.5,4)$ so $|A-B|^2=(.5-(-1.5))^2+(7-4)^2=13$.
       $|A-B|=\sqrt{13}$.

(d) $A=(\sqrt{2}/2,\sqrt{2}/2)$ and $B=(0,0)$ so $|A-B|^2=(\sqrt{2}/2-0)^2+(\sqrt{2}/2-0)^2=1$.
       $|A-B|=1$.

3.   Let $X=(x_1,x_2)$. Prove $|X|=|-X|$.

       $|X|^2=x_1^2+x_2^2=(-x_1)^2+(-x_2)^2=|-X|^2$. Therefore $|X|=|-X|$.

4.   (a)   Let $P=(p_1,p_2)$ and $Q=(q_1,q_2)$. $|P-Q|^2=(p_1-q_1)^2+(p_2-q_2)^2$.

       (b)   Let T_A be the translation by the point $A=(a_1,a_2)$.
              $T_A(P)=(p_1+a_1,p_2+a_2)$ and $T_A(Q)=(q_1+a_1,q_2+a_2)$.

       (c) $|T_A(P),T_A(Q)|^2=[(p_1+a_1)-(q_1+a_1)]^2+[(p_2+a_2)-(q_2+a_2)]^2$
              $=(p_1-q_1)^2+(p_2-q_2)^2=|P-Q|^2$. Therefore, $|P-Q|=|T_A(P)-T_A(Q)|$.
              Thus T_A is an isometry.

5.   To prove that reflection through the x-axis is an isometry, let $P=(a,b)$
       and $Q=(c,d)$ be points in the plane. Then
       $R_X(P)=(a,-b)$ and $R_X(Q)=(c,-d)$. We get
       $|R_X(P)-R_X(Q)|^2=(a-c)^2+(-b-(-d))^2$
       $\qquad\qquad\qquad\quad =(a-c)^2+[(-1)(b-d)]^2$
       $\qquad\qquad\qquad\quad =(a-c)^2+(b-d)^2=(P-Q|^2$
       Taking square roots we have $|P-Q|=|R_X(P)-R_X(Q)|$. Therefore, R_X is an
       isometry.

6. $d(-A,-B)=|(-B)-(-A)|$
       $\qquad\qquad =|A-B|$
       $\qquad\qquad =|A-B|$ by exercise 3
       $\qquad\qquad =d(a,B)$.
       Therefore, R_0 preserves distance.

§ 3. Exercises
(Pages 374-376)

1.   (a) $G_{30} \circ G_{65} = G_{95}$.

       (b) $G_{180} \circ G_{180} = G_0$.

       (c) $G_{170} \circ G_{225} = G_{35}$.

(d) $G_{980} \circ G_{100} = G_0$.

2.   (a) $T_{BC} \circ T_{AB}(P) = T_{BC}(P+B-A) = P+B-A+C-B = P+(C-A) = T_{AC}(P)$.

(b) $T_{BA} \circ T_{AB}(P) = T_{BA}(P+B-A) = P+B-A+A-B = P = I(P)$

(c) $T_{CB} \circ T_{AC}(P) = T_{CB}(P+C-A) = P+C-A+B-C = P+B-A = T_{AB}(P)$.

(d) $T_{CA} \circ T_{BC}(P) = T_{CA}(P+C-B) = P+C-B+A-C = P+A-B = T_{BA}(P)$.

3.

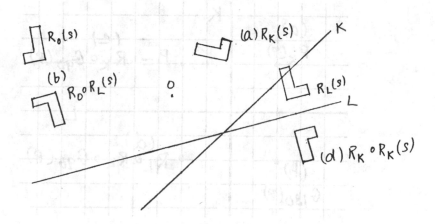

4. $G_{x,0} \circ G_{y,0} = G_{x+y,0} = G_{y+x,0} = G_{y,0} \circ G_{x,0}$.

$T_{AB} \circ T_{AC} = T_{AC} \circ T_{AB}$.

5. R_L and R_0 in exercise 3.
G_{90} and R_L on page 369.

6. $R_L \ o \ R_L = I.$
      $R_L \ o \ R_L \ o \ R_L = R_L.$

7.   If $T_{AB} \ o \ R_L = R_L \ o \ T_{AB}$ then the vector AB is parallel to line L.

8.

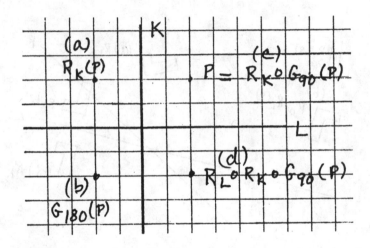

9.   (a) $T_{BC} \ o \ T_{AB} = T_{AC}$

      (b) $G_{145} \ 0 \ G_{35} = G_{180}$

      (c) $R_0 \ o \ G_{30} = G_{180} \ o \ G_{30} = G_{210}$

      (d) $R_0 \ o \ R_0 = I$

10. Given lines L and K intersecting at O and perpendicular, prove that $R_L \circ R_K = R_O$.

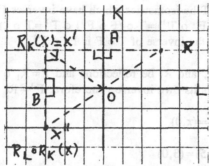

(a) Let X be a point in the plane. $R_K(X)=X'$ and $R_L \circ R_K(X)=X''$. By R_K we have $XX' \perp K$ at A and $|AX|=|AX'|$. By R_L we have $X''X' \perp L$ at B and $|BX'|=|BX''|$. Applying the RT Postulate to $\triangle AOX$ and $\triangle AOX'$ we get $|OX|=|OX'|$. Applying RT to $\triangle BOX'$ and $\triangle BOX''$ we get $|OX'|=|OX''|$. Thus $|OX|=|OX''|$. Since $m(\angle XOX')+m(\angle X'OX'')=180°$ $\angle XOX''$ is a straight angle and points X, O, and X'' are collinear. By definition of a reflection through a point, $R_O(X)=X''$. Therefore, $R_L \circ R_K = R_O$.

(b) Let L be the x-axis and K the y-axis and point O the origin. Let $X=(x_1,x_2)$. $R_y(X)=(-x_1,x_2)=X'$. $R_x(X')=(-x_1,-x_2)=R_O$. Therefore, $R_x \circ R_y = R_O$.

11. Let $A=(-3,4)$ and R be a reflection through the origin.

(a) (i) $R \circ T_A \circ R(5,6)=R \circ T_A(-5,-6)=R(-8,-2)=(8,2)$.

      (ii) $R \circ T_A \circ R(0,0)=R \circ T_A(0,0)=R(-3,4)=(3,-4)$.

      (iii) $R \circ T_A \circ R(-1,-1)=R \circ T_A(1,1)=R(-2,5)=(2,-5)$.

(b) $R \circ T_A \circ R$ does not equal T_A.
     $R \circ T_A \circ R(X)=R \circ T_A(-X)=R(-X+A)=(X-A)=T_{-A}(X)$.

12. Let $Q=(3,4)$ and R_Q be the reflection through Q.

(a) $P=(0,5)$, $R_Q(P)=(6,3)$. In the x direction $d_x(P,Q)=3$ so the image of P is 3 units beyond Q in the x direction. In the y direction $d_y(P,Q)=-1$ so the image of P is 1 unit below Q in the y direction.

(b) $P=(-1,2)$, $R_Q(P)=(7,6)$ Note: Q is the midpoint of PP'.

13. If $T=T^{-1}$ and T is a translation, T has $(0,0)$ as its vector and equals I.

14.   If $G=G^{-1}$ and G is a rotation, G could be I or G_{180} or equivalent rotation.

15.   Let F and T be isometries which have inverses F^{-1} and T^{-1}.
       (T o F) o $(F^{-1}$ o $T^{-1})$=T o (F o $F^{-1})$ o T^{-1}=T o I o T^{-1} =T o T^{-1}=I.
       Therefore, the inverse of T o F is F^{-1} o T^{-1}.

16.   Let R_1, R_2, and R_3 be reflections and $F=R_1$ o R_2 o R_3. Recall that a
       reflection is its own inverse. Therefore, $F^{-1}=R_3$ o R_2 o R_1.

17.   Let $F(x,y)=(x,0)$

       (a) $F(3,5)=(3,0)$

              $F(3,-2)=3,0)$

              $F(-6,0)=(-6,0)$

              $F(-6,10)=(-6,0)$

       (b)   F is not an isometry. For example, $d[(x,y),(0,0)]= \sqrt{x^2+y^2}$
              whereas $d[(x,0),(0,0)]=\sqrt{x^2}$. Thus the two distances are not equal
              if $y \neq 0$.

       (c)   There is no inverse. For example, $F(x,y)=F(x,0)=(x,0)$ since F
              maps all points on a vertical line to the same point on the x-axis.
              If there were an inverse T for F, then
              $(x,y)=T(F(x,y)=T(x,0)=T(F(x,0)=(x,0)$ and   this   is   a
              contradiction.

       (d)   Not every mapping has an inverse.

§ 4. Exercises
(Pages 379-381)

1.   (a), (g), and (h) are congruent. (c) and (e) are congruent. Finally (d) and
       (f) are congruent.

2.   A.   Reflect S through the y-axis.

       B.   Translate in the direction of the negative y-axis a distance equal to
             the height of S.

       C.   Rotate S by -45° around the origin.

D.   Translate as in part B, then reflect about the y-axis.

E.   Rotate 180° about the origin or reflect S through the origin.

3.   Let R_1 be a rectangle ABCD and R_2 be the rectangle MNOP. Let $|AB| = |MN|$, $|BC| = |NO|$, $|CD| = |OP|$, and $|DA| = |PM|$. Prove: $R_1 \cong R_2$

Let T be a translation that maps A to M. There is a rotation Q that maps AB to MN. Since $|AB| = |MN|$ Q o T maps B to N. Let Q(T(C))=C'. C' may be on the same side of MN as O. If not, reflect through line MN, R_{MN}. Q o T o R_{MN} (if needed) map BC to NO. These isometries also map D to P since BC maps to NO and angles are preserved. Thus, the composite isometry Q o T o R_{MN} (if needed) maps ABCD to MNOP and $R_1 \cong R_2$.

4.   For the kite quadrilateral ABCD let $|AB| = |BC|$ and $|AD| = |DC|$. Prove: $m(\angle a) = m(\angle C)$

Let R_{AB} be the reflection through the line $\overleftrightarrow{AB}$. Since $|AB| = |BC|$ and $|AD| = |DC|$, $R_{AB}(C) = A$. Since R_{AB} is an isometry $m(\angle A) = m(\angle C)$.

5.   Prove that two circles with the same radius, r, and centers O and O' are congruent.

Let T be the translation that maps O to O'. Let P be a point on the circle centered at O. T(P)=P'. Since T is an isometry, $|OP| = |O'P'| = r$. Therefore P' is on the circle centered at O'. Thus, T maps the circle centered at O with radius r to the circle centered at O' with radius r. These two circles are congruent by the definition of congruence.

§ 5. Exercise
(page 384)

1.   Given $\triangle ABC$ and $\triangle XYZ$, with $|AB| = |XY|$, $|AC| = |XZ|$, and $m(\angle A) = m(\angle X)$: let T be a translation that maps A to X. Rotate on X so that $\overline{A'B'}$ lies on $\overline{XY}$. Suppose C' is on the same side of $\overline{XY}$ as Z. Since these isometries preserve angle measure the image of the ray R_{AC} must coincide with the ray R_{XZ}. Furthermore, $|AC| = |XZ|$ so point C maps to point Z. Therefore, $\overline{BC}$ maps to $\overline{YZ}$. Thus, $\triangle ABC \cong \triangle XYZ$.

§ 6. Exercises
(Page 389)

1.   Theorem 12-7. Let F be an isometry, and suppose F has two distinct
        fixed points P and Q. Then every point of the line L_{PQ} is a fixed point.

        Proof. Let X be a point on L_{PQ}. Suppose first that X lies on the segment
        $\overline{PQ}$. Then $d(P,Q)=d(P,X)+d(X,Q)$.
        Since F is an isometry, we have
                        $d(P,X)=d(F(P),F(X))=d(P,F(X))$
                and $d(X,Q)=d(F(X),F(Q))=d(F(X),Q)$.
                Hence $d(P,Q)=d(P,F(X))+d(F(X),Q)$.
        By the SEG Postulate, F(X) lies on the segment $\overline{PQ}$ and, since distance
        is preserved F(X)=X.

        Now suppose X is on L_{PQ} but not on segment $\overline{PQ}$. Suppose X is on the
        ray R_{PQ}. Then, $d(P,X)=d(P,Q)+d(Q,X)$
        Since F is an isometry, we have $d(P,Q)=d(F(P),F(Q))=d(P,Q)$
        and $d(Q,X)=d(F(Q),F(X))=d(Q,F(X))$.
        Hence, $d(P,X)=d(P,Q)+d(Q,F(X))$.
        By the SEG Postulate, Q lies on the segment PF(X) and since
        $d(Q,X)=d(Q,F(X))$, F(X)=X.
        This concludes the proof.

2.   Let F, T be isometries. Let P, Q, M be three points not on a straight line
        such that F(P)=T(P), F(Q)=T(Q), F(M)=T(M).
        Prove; F=T.

        Let G be the inverse of F. Then $G(T(P))=P$, $G(T(Q))=Q$, and
        $G(T(M))=M$. By Theorem 12.8, if an isometry has three distinct fixed
        points, not lying on the same line, then that isometry is the identity. Thus
        $G \circ T = I$. Since G is the inverse of F, $F \circ G = I$. Thus,
        $T = I \circ T = (F \circ G) \circ T = F \circ (G \circ T) = F \circ I = F$.

3.   Let F be an isometry with one fixed point.
        Prove: there is a rotation G such that G o F has two fixed points.

        Let P be the fixed point of F such that F(P)=P. Let X be any other point
        and let F(X)=X'. If X=X' we are done. Suppose X is not X'. Since F
        is an isometry, $d(P,X)=d(F(P),F(X))=d(P,X')$. Thus X and X' lie on a
        circle centered at P with radius d(P,X). Let G be the rotation centered at

P that maps X' to X.
Then G o F(X)=X and G o F(P)=P, so G o F has two fixed points.

4.   Let F be an isometry. Prove that there is a translation T such that T o F
has a fixed point.

If F is the identity, we are done.
If F is not the identity, let P be a point such that $F(P)=P'$ and P is not
P'. Define T to be the translation $T_{P'P}$.
Then $T_{P'P}$ o $F(P)=T_{P'P}(P')=P$ so P is a fixed point for $T_{P'P}$ o F.

5.   Prove that every isometry F can be written in the form
        F=T o G o R or F=T o G or F=G or F=I
where R is a reflection through a line, G is a rotation, and T is a
translation.

By Exercise 4 there exists T such that T o F has a fixed point. By
Exercise 3 there is a rotation G such that G o T o F has two fixed points,
P and Q. By Theorem 12-11 either G o T o F=I or G o T o $F=R_L$
where $L=L_{PQ}$. If G o T o F=I, then $F=T^{-1}$ o G^{-1}. If G o T o F=R, then
$F= T^{-1}$ o G^{-1} o R. If F has one fixed point, then F=G. If F has 3 fixed
points, F=I.

6.   Prove: Every isometry has an inverse.
By Theorem 12-12, F is the identity or F is a composition of at most
three reflections. If F is the identity, I, I is its own inverse. Suppose F
is not I. By Exercise 16 in 12.3, if $F=R_1$ o R_2 o R_3, then $F^{-1}=R_3$ o R_2
o R_1. Similarly, $(R_1$ o $R_2)^{-1}= R_2$ o R_1 and $R^{-1}=R$. Thus, every isometry
has an inverse.